Real-Time Embedded Multithreading
Using ThreadX and MIPS

T0227610

Real-Time Embedded Multithreading Using ThreadX and MIPS

Edward L. Lamie

Routledge
Taylor & Francis Group

LONDON AND NEW YORK

First published 2009 by Elsevier Inc.

Published 2018 by Routledge
2 Park Square, Milton Park, Abingdon, Oxon OX14 4RN
52 Vanderbilt Avenue, New York, NY 10017

Routledge is an imprint of the Taylor & Francis Group, an informa business

Library of Congress Cataloging-in-Publication Data
Application submitted

British Library Cataloguing-in-Publication Data
A catalogue record for this book is available from the British Library.

Typeset by Charon Tec Ltd., A Macmillan Company. (www.macmillansolutions.com)

ISBN 13: 978-1-85617-631-6 (pbk)
ISBN 13: 978-1-138-46079-9 (hbk)

à mes ancêtres québecois

Contents

Preface

Embedded systems are ubiquitous. These systems are found in most consumer electronics, automotive, government, military, communications, and medical equipment. Most individuals in developed countries have many such systems and use them daily, but relatively few people realize that these systems actually contain embedded computer systems. Although the field of embedded systems is young, the use and importance of these systems is increasing, and the field is rapidly growing and maturing.

This book is intended for persons who develop embedded systems, or for those who would like to know more about the process of developing such systems. Although embedded systems developers are typically software engineers or electrical engineers, many people from other disciplines have made significant contributions to this field. This book is specifically targeted toward embedded applications that must be small, fast, reliable, and deterministic.[1]

This book is composed of 14 chapters that cover embedded and real-time concepts, the MIPS® processor, all the services provided by the ThreadX® real-time operating system (RTOS), solutions to classical problem areas, and a case study. I assume the reader has a programming background in C or C++, so we won't devote any time to programming fundamentals. Depending on the background of the reader, the chapters of the book may be read independently.

There are several excellent books written about embedded systems. However, most of these books are written from a generalist point of view. This book is unique because it is based on embedded systems development using a typical commercial RTOS, as well as a typical microprocessor. This approach has the advantage of providing specific knowledge and techniques, rather than generic concepts that must be converted to your specific system. Thus, you can immediately apply the topics in this book to your development efforts.

Because an actual RTOS is used as the primary tool for embedded application development, there is no discussion about the merits of building your own RTOS or

---

[1]Such systems are sometimes called deeply embedded systems.

forgoing an RTOS altogether. I believe that the relatively modest cost of a commercial RTOS provides a number of significant advantages over attempts to "build your own." For example, most commercial RTOS companies have spent years refining and optimizing their systems. Their expertise and product support may play an important role in the successful development of your system.

The RTOS chosen for use in this book is ThreadX[2] (version 5). This RTOS was selected for a variety of reasons, including reliability, ease of use, low cost, widespread use, and the maturity of the product due to the extensive experience of its developers. This RTOS contains most of the features found in contemporary RTOSes, as well as several advanced features that are not. Another notable feature of this RTOS is the consistent and readable coding convention used within its application programming interface (API). Developing applications is highly intuitive because of the logical approach of the API.

Although I chose the C programming language for this book, you could use C++ instead for any of the applications described in this book.

There is a CD included with this book that contains a limited ThreadX[3] system. You may use this system to perform your own experiments, run the included demonstration system, and experiment with the projects described throughout the book.

Typographical conventions are used throughout this book so that key concepts are communicated easily and unambiguously. For example, keywords such as main or int are displayed in a distinctive typeface, whether these keywords are in a program or appear in the discussion about a program. This typeface is also used for all program segment listings or when actual input or output is illustrated. When an identifier name such as *MyVar* is used in the narrative portion of the book, it will appear in italics. The italics typeface will also be used when new topics are introduced or to provide emphasis.

---

[2]ThreadX is a registered trademark of Express Logic, Inc. The ThreadX API, associated data structures, and data types are copyrights of Express Logic, Inc. MIPS is a registered trademark of MIPS Processors, Inc.

[3]Express Logic, Inc. has granted permission to use this demonstration system for the sample systems and the case study in this book.

CHAPTER 1

Embedded and Real-time Systems

1.1 Introduction

Although the history of embedded systems is relatively short,[1] the advances and successes of this field have been profound. Embedded systems are found in a vast array of applications such as consumer electronics, "smart" devices, communication equipment, automobiles, desktop computers, and medical equipment.[2]

1.2 What is an Embedded System?

In recent years, the line between embedded and nonembedded systems has blurred, largely because embedded systems have expanded to a vast array of applications. However, for practical purposes, an embedded system is defined here as one dedicated to a specific purpose and consisting of a compact, fast, and extremely reliable operating system that controls the microprocessor located inside a device. Included in the embedded system is a collection of programs that run under that operating system, and of course, the microprocessor.[3]

---

[1]The first embedded system was developed in 1971 by the Intel Corporation, which produced the 4004 microprocessor chip for a variety of business calculators. The same chip was used for all the calculators, but software in ROM provided unique functionality for each calculator. Source: The Intel 4004 website at http://www.intel4004.com/

[2]Approximately 98% of all microprocessors are used in embedded systems. Turley, Jim, *The Two Percent Solution*, Embedded Systems Programming, Vol. 16, No. 1, January 2003.

[3]The microprocessor is often called a *microcontroller, embedded microcontroller, network processor,* or *digital signal processor*; it consists of a CPU, RAM, ROM, I/O ports, and timers.

Because an embedded system is part of a larger system or device, it is typically housed on a single microprocessor board and the associated programs are stored in ROM.[4] Because most embedded systems must respond to inputs within a small period of time, these systems are frequently classified as real-time systems. For simple applications, it might be possible for a single program (without an RTOS) to control an embedded system, but typically an RTOS or kernel is used as the engine to control the embedded system.

1.3 Characteristics of Embedded Systems

Another important feature of embedded systems is *determinism*. There are several aspects to this concept, but each is built on the assumption that for each possible state and each set of inputs, a unique set of outputs and next state of the system can be, in principle, predicted. This kind of determinism is not unique to embedded systems; it is the basis for virtually all kinds of computing systems. When you say that an embedded system is deterministic, you are usually referring to *temporal determinism*. A system exhibits temporal determinism if the time required to process any task is finite and predictable. In particular, we are less concerned with average response time than we are with worst-case response time. In the latter case, we must have a guarantee on the upper time limit, which is an example of temporal determinism.

An embedded system is typically encapsulated by the hardware it controls, so end-users are usually unaware of its presence. Thus, an embedded system is actually a computer system that does not have the outward appearances of a computer system. An embedded system typically interacts with the external world, but it usually has a primitive or nonexistent user interface.

The embedded systems field is a hybrid that draws extensively from disciplines such as software engineering, operating systems, and electrical engineering. Embedded systems has borrowed liberally from other disciplines and has adapted, refined, and enhanced those concepts and techniques for use in this relatively young field.

1.4 Real-time Systems

As noted above, an embedded system typically must operate within specified time constraints. When such constraints exist, we call the embedded system a *real-time system*.

---

[4]We often say that embedded systems are *ROMable* or *scalable*.

This means that the system must respond to inputs or events within prescribed time limits, and the system as a whole must operate within specified time constraints. Thus, a real-time system must not only produce correct results, but also it must produce them in a timely fashion. The timing of the results is sometimes as important as their correctness.

There are two important subclasses of real-time constraints: hard real-time and soft real-time. Hard real-time refers to highly critical time constraints in which missing even one time deadline is unacceptable, possibly because it would result in catastrophic system failure. Examples of hard real-time systems include air traffic control systems, medical monitoring systems, and missile guidance systems. Soft real-time refers to situations in which meeting the time constraints is desirable, but not critical to the operation of the system.

1.5 Real-time Operating Systems and Real-time Kernels

Relatively few embedded applications can be developed effectively as a single control program, so we consider only commercially available real-time operating systems (RTOSes) and real-time kernels here. A real-time kernel is generally much smaller than a complete RTOS. In contemporary operating system terminology, a *kernel* is the part of the operating system that is loaded into memory first and remains in memory while the application is active. Likewise, a real-time kernel is memory-resident and provides all the necessary services for the embedded application. Because it is memory-resident, a real-time kernel must be as small as possible. Figure 1.1 contains an illustration of a typical kernel and other RTOS services.

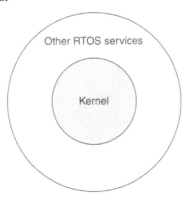

Figure 1.1: RTOS kernel

The operation of an embedded system entails the execution of processes, and tasks or threads, either in response to external or internal inputs, or in the normal processing required for that system. The processing of these entities must produce correct results within specified time constraints.

1.6 Processes, Tasks, and Threads

The term *process* is an operating system concept that refers to an independent executable program that has its own memory space. The terms "process" and "program" are often used synonymously, but technically a process is more than a program: it includes the execution environment for the program and handles program bookkeeping details for the operating system. A process can be launched as a separately loadable program, or it can be a memory-resident program that is launched by another process. Operating systems are often capable of running many processes concurrently. Typically, when an operating system executes a program, it creates a new process for it and maintains within that process all the bookkeeping information needed. This implies that there is a one-to-one relationship between the program and the process, i.e., one program, one process.

When a program is divided into several segments that can execute concurrently, we refer to these segments as threads. A *thread* is a semi-independent program segment; threads share the same memory space within a program. The terms "task" and "thread" are frequently used interchangeably. However, we will use the term "thread" in this book because it is more descriptive and more accurately reflects the processing that occurs. Figure 1.2 contains an illustration of the distinction between processes and threads.

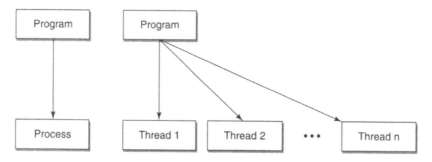

Figure 1.2: Comparison of processes and threads

1.7 Architecture of Real-time Systems

The architecture of a real-time system determines how and when threads are processed. Two common architectures are the *control loop with polling*[5] approach and the *preemptive scheduling* model. In the control loop with polling approach, the kernel executes an infinite loop, which polls the threads in a predetermined pattern. If a thread needs service, then it is processed. There are several variants to this approach, including *time-slicing*[6] to ensure that each thread is guaranteed access to the processor. Figure 1.3 contains an illustration of the control loop with polling approach.

Although the control loop with polling approach is relatively easy to implement, it has several serious limitations. For example, it wastes much time because the processor polls threads that do not need servicing, and a thread that needs attention has to wait its turn until the processor finishes polling other threads. Furthermore, this approach makes no

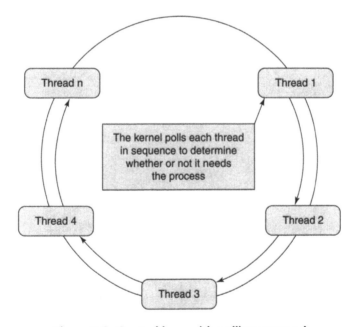

Figure 1.3: Control loop with polling approach

---

[5]The control loop with polling approach is sometimes called the *super loop* approach.
[6]Each thread is allocated a predetermined slice of time in which to execute.

distinction between the relative importance of the threads, so it is difficult to give threads with critical requirements fast access to the processor.

Another approach that real-time kernels frequently use is *preemptive scheduling*. In this approach, threads are assigned priorities and the kernel schedules processor access for the thread with the highest priority. There are several variants to this approach including techniques to ensure that threads with lower priorities get some access to the processor. Figure 1.4 illustrates one possible implementation of this approach. In this example, each thread is assigned a priority from zero (0) to some upper limit.[7] Assume that priority zero is the highest priority.

An essential feature in preemptive scheduling schemes is the ability to suspend the processing of a thread when a thread that has a higher priority is ready for processing. The process of saving the current information of the suspended thread so that another thread can execute is called *context switching*. This process must be fast and reliable

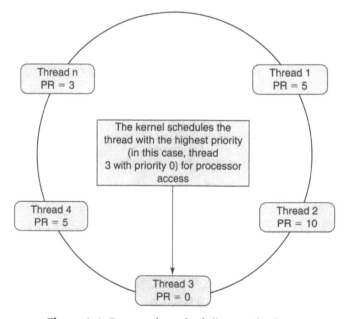

Figure 1.4: Preemptive scheduling method

[7]ThreadX provides 1024 distinct priority values, where 0 represents the highest priority.

because the suspended thread must be able to resume execution exactly at the point where it was suspended when it ultimately regains control of the processor.

Embedded systems need to respond to inputs or events accurately and within specified deadlines. This is accomplished in part by means of an *interrupt*, which is a signal to the processor that an event has occurred and that immediate attention may be required. An interrupt is handled with an *interrupt service routine (ISR)*, which may activate a thread with a higher priority than the currently executing thread. In this case, the ISR would suspend the currently executing thread and permit the higher priority thread to proceed. Interrupts can be generated from software[8] or by a variety of hardware devices.

1.8 Embedded Systems Development

Embedded applications should be designed and developed using sound software engineering principles. Because most embedded applications are real-time systems, one major difference from traditional computer applications is the requirement to adhere strictly to prescribed time constraints.[9] The requirements and design phases are performed with the same rigor as any other software application.

Another major consideration in embedded systems development is that the modules (that is, the threads) are not designed to be executed in a procedural manner, as is the case with traditional software systems. The threads of an embedded application are designed to be executed independently of each other or in parallel[10] so this type of system is called *multithreaded*.[11] Because of this apparent parallelism, the traditional software-control structures are not always applicable to embedded systems.

A real-time kernel is used as the engine to drive the embedded application, and the software design consists of threads to perform specific operations, using inter-thread communication facilities provided by the kernel. Although most embedded systems development is done in the C (or C++) programming language, some highly critical portions of the application are often developed in assembly language.

---

[8]Software interrupts are also called *traps* or *exceptions*.

[9]Some writers liken the study of real-time systems to the science of *performance guarantees*.

[10]In cases where there is only one processor, threads are executed in *pseudo-parallel*.

[11]Multithreading is sometimes called *multitasking*.

1.9 Key Terms and Phrases

control loop with polling	priority
determinism	real-time kernel
embedded system	real-time system
interrupt	ROMable
microprocessor	RTOS
multithreading	scalable
preemptive scheduling	Thread

CHAPTER 2

First Look at a System Using an RTOS

2.1 Operating Environment

We will use the Win32 version of ThreadX because it permits developers to develop prototypes of their applications in the easy-to-use and prevalent Windows programming environment. We achieve complete ThreadX simulation by using Win32 calls. The ThreadX-specific application code developed in this environment will execute in an identical fashion on the eventual target hardware. Thus, ThreadX simulation allows real software development to start well before the actual target hardware is available. We will use Microsoft Visual C/C++ Tools to compile all the embedded systems in this book.

2.2 Installation of the ThreadX Demonstration System

There is a demonstration version of ThreadX on the CD included with this book. View the Readme file for information about installing and using this demonstration system.

2.3 Sample System with Two Threads

The first step in mastering the use of ThreadX is to understand the nature and behavior of threads. We will achieve this purpose by performing the following operations in this sample system: create several threads, assign several activities to each thread, and compel the threads to cooperate in the execution of their activities. A *mutex* will be used to coordinate the thread activities, and a memory byte pool will be used to create *stacks* for the threads. (Mutexes and stacks are described in more detail later.)

The first two components that we create are two threads named Speedy_Thread and Slow_Thread. Speedy_Thread will have a higher priority than Slow_Thread and will generally finish its activities more quickly. ThreadX uses a preemptive scheduling algorithm, which means that threads with higher priorities generally have the ability to preempt the execution of threads with lower priorities. This feature may help Speedy_Thread to complete its activities more quickly than Slow_Thread. Figure 2.1 contains an illustration of the components that we will use in the sample system.

In order to create the threads, you need to assign each of them a *stack*: a place where the thread can store information, such as return addresses and local variables, when it is preempted. Each stack requires a block of contiguous bytes. You will allocate these bytes from a memory byte pool, which you will also create. The memory byte pool could also be used for other ThreadX objects, but we will restrict its usage to the two threads in this system. There are other methods by which we could assign memory space for a stack, including use of an array and a memory block pool (to be discussed later). We choose to use the memory byte pool in this sample system only because of its inherent simplicity.

We will use a ThreadX object called a mutex in this sample system to illustrate the concept of mutual exclusion. Each of the two threads has two sections of code known as *critical sections*. Very generally, a critical section is one that imposes certain constraints on thread execution. In the context of this example, the constraint is that when a thread is executing a critical section, it must not be preempted by any other thread executing a critical section—no two threads can be in their respective critical sections at the same

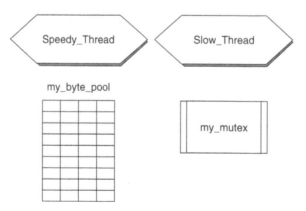

Figure 2.1: Components of the sample system

time. A critical section typically contains shared resources,[1] so there is the potential for system failure or unpredictable behavior when more than one thread is in a critical section.

A *mutex* is an object that acts like a token or gatekeeper. To gain access to a critical section, a thread must acquire "ownership" of the mutex, and only one thread can own a given mutex at the same time. We will use this property to provide inter-thread mutual exclusion protection. For example, if Slow_Thread owns the mutex, then Speedy_Thread must wait to enter a critical section until Slow_Thread gives up ownership of the mutex, even though Speedy_Thread has a higher priority. Once a thread acquires ownership of a mutex, it will retain ownership until it voluntarily gives up that mutex. In other words, no thread can preempt a mutex owned by another thread regardless of either thread's priority. This is an important feature that provides inter-thread mutual exclusion.

Each of the two threads in the sample system has four activities that will be executed repeatedly. Figure 2.2 contains an illustration of the activities for the Speedy_Thread. Activities 2 and 4 appear in shaded boxes that represent critical sections for that thread. Similarly, Figure 2.3 contains an illustration of the activities for the Slow_Thread. Note that Speedy_Thread has a priority of 5, which is higher than the priority of 15 that is assigned to the Slow_Thread.

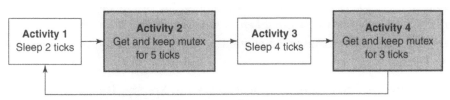

Figure 2.2: Activities of the Speedy_Thread (priority = 5)

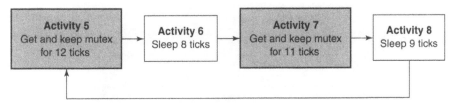

Figure 2.3: Activities of the Slow_Thread (priority = 15)

---

[1]Or, it contains code that accesses shared resources.

2.4 Creating the ThreadX Objects

Program listing *02_sample_system.c* is located at the end of this chapter and on the attached CD. It contains the complete source code for our sample system. Detailed discussion of the specifics of this listing is included in later chapters to provide a highlight of the essential portions of the system. Figure 2.4 contains a summary of the main features of the source code listing.

The main() portion of the basic structure contains exactly one executable statement, as follows:

tx_kernel_enter();

The above entry function turns over control to ThreadX (and does not return!). ThreadX performs initialization of various internal data structures and then processes the application definitions and the thread entry definitions. ThreadX then begins scheduling and executing application threads. The purpose of the tx_application_define function in our sample system is to define all the ThreadX components that will be used. For example, we need to define a memory byte pool, two threads, and one mutex. We also need to allocate memory from the byte pool for use as thread stacks. The purpose of the thread entry functions section is to prescribe the behavior of the two threads in the system. We will consider only one of the *thread entry functions* in this discussion because both entry functions are similar. Figure 2.5 contains a listing of the entry function for the Speedy_Thread.

Recall that activities 2 and 4 are the critical sections of Speedy_Thread. Speedy_Thread seeks to obtain ownership of the mutex with the following statement:

tx_mutex_get(&my_mutex, TX_WAIT_FOREVER);

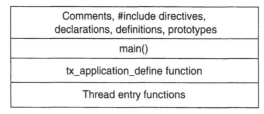

Figure 2.4: Basic structure of sample system

If Slow_Thread already owns the mutex, then Speedy_Thread will "wait forever" for its turn to obtain ownership. When Speedy_Thread completes a critical section, it gives up ownership of the mutex with the following statement:

tx_mutex_put(&my_mutex);

When this statement is executed, Speedy_Thread relinquishes ownership of the mutex, so it is once again available. If Slow_Thread is waiting for the mutex, it will then have the opportunity to acquire it.

```
/* Entry function definition of the "Speedy_Thread"
   which has a higher priority than the "Slow_Thread" */

void  Speedy_Thread_entry(ULONG thread_input)
{
UINT   status;
ULONG current_time;

   While (1)
   {
       /* Activity 1:  2 timer-ticks */
       tx_thread_sleep(2);

       /* Get the mutex with suspension */
       tx_mutex_get(&my_mutex, TX_WAIT_FOREVER);

       /* Activity 2:  5 timer-ticks */ /* *** critical section *** */
       tx_thread_sleep(5);

       /* Release the mutex */
       tx_mutex_put(&my_mutex);

       /* Activity 3:  4 timer-ticks */
       tx_thread_sleep(4);

       /* Get the mutex with suspension */
       tx_mutex_get(&my_mutex, TX_WAIT_FOREVER);

       /* Activity 4:  3 timer-ticks  *** critical section *** */
       tx_thread_sleep(3);

       /* Release the mutex */
       tx_mutex_put(&my_mutex);

       current_time = tx_time_get();
       printf("Current Time: %5lu  Speedy_Thread finished cycle...\n",
               current_time);
   }
}
```

Figure 2.5: Entry function definition for the Speedy_Thread

The entry function for Speedy_Thread concludes by getting the current system time and displaying that time along with a message that Speedy_Thread has finished its current cycle of activities.

2.5 Compiling and Executing the Sample System

Compile and execute the sample system contained in *02_sample_system.c* that is located on the attached CD. A complete listing appears in a section at the end of this chapter.

2.6 Analysis of the System and the Resulting Output

Figure 2.6 contains output produced by executing the sample system. Your output should be similar, but not necessarily identical.

The minimum amount of time in which Speedy_Thread can complete its cycle of activities is 14 timer-ticks. By contrast, the Slow_Thread requires at least 40 timer-ticks to complete one cycle of its activities. However, the critical sections of the Slow_Thread will cause delays for the Speedy_Thread. Consider the sample output in Figure 2.6 where the Speedy_Thread finishes its first cycle at time 34, meaning that it encountered a delay of 20 timer-ticks because of the Slow_Thread. The Speedy_Thread completes subsequent cycles in a more timely fashion but it will always spend a lot of time waiting for the Slow_Thread to complete its critical section.

2.7 Listing of 02_sample_system.c

The sample system named *02_sample_system.c* is located on the attached CD. The complete listing appears below; line numbers have been added for easy reference.

```
Current Time:     34   Speedy_Thread finished cycle...
Current Time:     40   Slow_Thread finished cycle...
Current Time:     56   Speedy_Thread finished cycle...
Current Time:     77   Speedy_Thread finished cycle...
Current Time:     83   Slow_Thread finished cycle...
Current Time:     99   Speedy_Thread finished cycle...
Current Time:    120   Speedy_Thread finished cycle...
Current Time:    126   Slow_Thread finished cycle...
Current Time:    142   Speedy_Thread finished cycle...
Current Time:    163   Speedy_Thread finished cycle...
```

Figure 2.6: Output produced by sample system

```
001  /* 02_sample_system.c
002
003  Create two threads, one byte pool, and one mutex.
004  The threads cooperate with each other via the mutex. */
005
006
007  /****************************************************/
008  /*      Declarations, Definitions, and Prototypes    */
009  /****************************************************/
010
011  #include "tx_api.h"
012  #include <stdio.h >
013
014  #define DEMO_STACK_SIZE      1024
015  #define DEMO_BYTE_POOL_SIZE 9120
016
017
018  /* Define the ThreadX object control blocks… */
019
020  TX_THREAD     Speedy_Thread;
021  TX_THREAD     Slow_Thread;
022
023  TX_MUTEX      my_mutex;
024
025  TX_BYTE_POOL my_byte_pool;
026
027
028  /* Define thread prototypes. */
029
030  void    Speedy_Thread_entry(ULONG thread_input);
031  void    Slow_Thread_entry(ULONG thread_input);
032
033
034  /****************************************************/
035  /*              Main Entry Point                     */
036  /****************************************************/
037
038  /* Define main entry point. */
039
040  int main()
041  {
```

```
042
043            /* Enter the ThreadX kernel. */
044            tx_kernel_enter();
045    }
046
047
048
049    /**************************************************/
050    /*              Application Definitions           */
051    /**************************************************/
052
053
054    /* Define what the initial system looks like. */
055
056    void tx_application_define(void *first_unused_memory)
057    {
058
059    CHAR *pool_pointer;
060
061
062    /* Create a byte memory pool from which to allocate
063       the thread stacks. */
064    tx_byte_pool_create(&my_byte_pool, "my_byte_pool",
065                        first_unused_memory,
066                        DEMO_BYTE_POOL_SIZE);
067
068    /* Put system definition stuff in here, e.g., thread
069       creates and other assorted create information. */
070
071    /* Allocate the stack for the Speedy_Thread. */
072    tx_byte_allocate(&my_byte_pool, (VOID **) &pool_pointer,
073                     DEMO_STACK_SIZE, TX_NO_WAIT);
074
075    /* Create the Speedy_Thread. */
076    tx_thread_create(&Speedy_Thread, "Speedy_Thread",
077                     Speedy_Thread_entry, 0,
078                     pool_pointer, DEMO_STACK_SIZE, 5, 5,
079                     TX_NO_TIME_SLICE, TX_AUTO_START);
080
081    /* Allocate the stack for the Slow_Thread. */
```

```
082   tx_byte_allocate(&my_byte_pool, (VOID **) &pool_pointer,
083                   DEMO_STACK_SIZE, TX_NO_WAIT);
084
085   /* Create the Slow_Thread. */
086   tx_thread_create(&Slow_Thread, "Slow_Thread",
087                    Slow_Thread_entry, 1, pool_pointer,
088                    DEMO_STACK_SIZE, 15, 15,
089                    TX_NO_TIME_SLICE, TX_AUTO_START);
090
091   /* Create the mutex used by both threads */
092   tx_mutex_create(&my_mutex, "my_mutex", TX_NO_INHERIT);
093
094
095   }
096
097
098   /****************************************************/
099   /*                 Function Definitions             */
100   /****************************************************/
101
102
103   /* Entry function definition of the "Speedy_Thread"
104      it has a higher priority than the "Slow_Thread" */
105
106   void Speedy_Thread_entry(ULONG thread_input)
107   {
108
109   ULONG current_time;
110
111    while (1)
112    {
113        /* Activity 1: 2 timer-ticks */
114        tx_thread_sleep(2);
115
116        /* Get the mutex with suspension */
117        tx_mutex_get(&my_mutex, TX_WAIT_FOREVER);
118
119        /* Activity 2: 5 timer-ticks *** critical section *** */
120        tx_thread_sleep(5);
121
```

```
122            /* Release the mutex */
123            tx_mutex_put(&my_mutex);
124
125            /* Activity 3: 4 timer-ticks */
126            tx_thread_sleep(4);
127
128            /* Get the mutex with suspension */
129            tx_mutex_get(&my_mutex, TX_WAIT_FOREVER);
130
131            /* Activity 4: 3 timer-ticks *** critical section *** */
132            tx_thread_sleep(3);
133
134            /* Release the mutex */
135            tx_mutex_put(&my_mutex);
136
137            current_time = tx_time_get();
138            printf("Current Time: %5lu Speedy_Thread finished a
                      cycle...\n",
139                   current_time);
140
141    }
142  }
143
144  /****************************************************/
145
146  /* Entry function definition of the "Slow_Thread"
147     it has a lower priority than the "Speedy_Thread" */
148
149  void Slow_Thread_entry(ULONG thread_input)
150  {
151
152
153  ULONG current_time;
154
155    while(1)
156    {
157        /* Activity 5 - 12 timer-ticks *** critical section *** */
158
159        /* Get the mutex with suspension */
160        tx_mutex_get(&my_mutex, TX_WAIT_FOREVER);
161
```

```
162          tx_thread_sleep(12);
163
164          /* Release the mutex */
165          tx_mutex_put(&my_mutex);
166
167          /* Activity 6 - 8 timer-ticks */
168          tx_thread_sleep(8);
169
170           /* Activity 7 - 11 timer-ticks *** critical section *** */
171
172          /* Get the mutex with suspension */
173          tx_mutex_get(&my_mutex, TX_WAIT_FOREVER);
174
175          tx_thread_sleep(11);
176
177          /* Release the mutex */
178          tx_mutex_put(&my_mutex);
179
180          /* Activity 8 - 9 timer-ticks */
181          tx_thread_sleep(9);
182
183          current_time = tx_time_get();
184          printf("Current Time: %5lu Slow_Thread finished a cycle...\n",
185                  current_time);
186
187   }
188 }
```

2.8 Key Terms and Phrases

application define function	preemption
critical section	priority
current time	scheduling threads
initialization	sleep time
inter-thread mutual exclusion	stack
kernel entry	suspension
memory byte pool	template
mutex	thread
mutual exclusion	thread entry function
ownership of mutex	timer-tick

2.9 Problems

1. Modify the sample system to compute the average cycle time for the Speedy Thread and the Slow Thread. You will need to add several variables and perform several computations in each of the two thread entry functions. You will also need to get the current time at the beginning of each thread cycle.

2. Modify the sample system to bias it in favor of the Speedy Thread. For example, ensure that Slow Thread will not enter a critical section if the Speedy Thread is within two timer-ticks of entering its critical section. In that case, the Slow Thread would sleep two more timer-ticks and then attempt to enter its critical section.

CHAPTER 3

RTOS Concepts and Definitions

3.1 Introduction

The purpose of this chapter is to review some of the essential concepts and definitions used in embedded systems. You have already encountered several of these terms in previous chapters, and you will read about several new concepts here.

3.2 Priorities

Most embedded real-time systems use a priority system as a means of establishing the relative importance of threads in the system. There are two classes of priorities: static and dynamic. A *static priority* is one that is assigned when a thread is created and remains constant throughout execution. A *dynamic priority* is one that is assigned when a thread is created, but can be changed at any time during execution. Furthermore, there is no limit on the number of priority changes that can occur.

ThreadX provides a flexible method of dynamic priority assignment. Although each thread must have a priority, ThreadX places no restrictions on how priorities may be used. As an extreme case, all threads could be assigned the same priority that would never change. However, in most cases, priority values are carefully assigned and modified only to reflect the change of importance in the processing of threads. As illustrated by Figure 3.1, ThreadX provides priority values from 0 to 31, inclusive, where the value 0 represents the highest priority and the value 31 represents the lowest priority.[1]

---

[1]The default priority range for ThreadX is 0 through 31, but up to 1024 priority levels can be used.

Priority value	Meaning
0	Highest priority
1	
:	
31	Lowest priority

Figure 3.1: Priority values

3.3 Ready Threads and Suspended Threads

ThreadX maintains several internal data structures to manage threads in their various states of execution. Among these data structures are the Suspended Thread List and the Ready Thread List. As implied by the nomenclature, threads on the Suspended Thread List have been *suspended*—temporarily stopped executing—for some reason. Threads on the Ready Thread List are not currently executing but are ready to run.

When a thread is placed in the Suspended Thread List, it is because of some event or circumstance, such as being forced to wait for an unavailable resource. Such a thread remains in that list until that event or circumstance has been resolved. When a thread is removed from the Suspended Thread List, one of two possible actions occurs: it is placed on the Ready Thread List, or it is terminated.

When a thread is ready for execution, it is placed on the Ready Thread List. When ThreadX schedules a thread for execution, it selects and removes the thread in that list that has the highest priority. If all the threads on the list have equal priority, ThreadX selects the thread that has been waiting the longest.[2] Figure 3.2 contains an illustration of how the Ready Thread List appears.

If for any reason a thread is not ready for execution, it is placed in the Suspended Thread List. For example, if a thread is waiting for a resource, if it is in "sleep" mode, if it was

---

[2]This latter selection algorithm is commonly known as First In First Out, or FIFO.

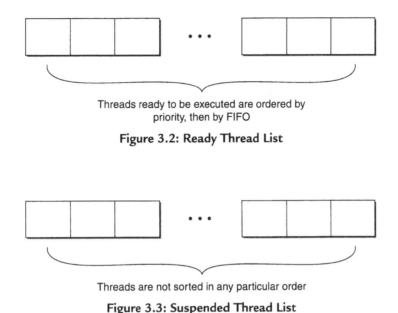

Threads ready to be executed are ordered by
priority, then by FIFO

Figure 3.2: Ready Thread List

Threads are not sorted in any particular order

Figure 3.3: Suspended Thread List

created with a TX_DONT_START option, or if it was explicitly suspended, then it will reside in the Suspended Thread List until that situation has cleared. Figure 3.3 contains a depiction of this list.

3.4 Preemptive, Priority-Based Scheduling

The term *preemptive, priority-based scheduling* refers to the type of scheduling in which a higher priority thread can interrupt and suspend a currently executing thread that has a lower priority. Figure 3.4 contains an example of how this scheduling might occur.

In this example, Thread 1 has control of the processor. However, Thread 2 has a higher priority and becomes ready for execution. ThreadX then interrupts Thread 1 and gives Thread 2 control of the processor. When Thread 2 completes its work, ThreadX returns control to Thread 1 at the point where it was interrupted. The developer does not have to be concerned about the details of the scheduling process. Thus, the developer is able to develop the threads in isolation from one another because the scheduler determines when to execute (or interrupt) each thread.

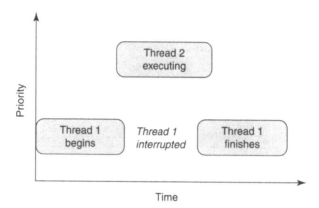

Figure 3.4: Thread preemption

3.5 Round-Robin Scheduling

The term *round-robin scheduling* refers to a scheduling algorithm designed to provide processor sharing in the case in which multiple threads have the same priority. There are two primary ways to achieve this purpose, both of which are supported by ThreadX.

Figure 3.5 illustrates the first method of round-robin scheduling, in which Thread 1 is executed for a specified period of time, then Thread 2, then Thread 3, and so on to Thread *n*, after which the process repeats. See the section titled *Time-Slice* for more information about this method. The second method of round-robin scheduling is achieved by the use of a cooperative call made by the currently executing thread that temporarily relinquishes control of the processor, thus permitting the execution of other threads of the same or higher priority. This second method is sometimes called *cooperative multithreading*. Figure 3.6 illustrates this second method of round-robin scheduling.

With cooperative multithreading, when an executing thread relinquishes control of the processor, it is placed at the end of the Ready Thread List, as indicated by the shaded thread in the figure. The thread at the front of the list is then executed, followed by the next thread on the list, and so on until the shaded thread is at the front of the list. For convenience, Figure 3.6 shows only ready threads with the same priority. However, the Ready Thread List can hold threads with several different priorities. In that case, the scheduler will restrict its attention to the threads that have the highest priority.

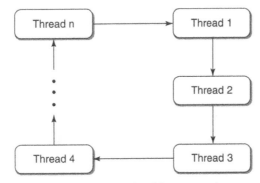

Figure 3.5: Round-robin processing

Ready thread list containing threads with the same
priority currently executing thread (shaded) voluntarily relinquishes
the processor and is placed on this list.

Figure 3.6: Example of cooperative multithreading

In summary, the cooperative multithreading feature permits the currently executing thread to voluntarily give up control of the processor. That thread is then placed on the Ready Thread List and it will not gain access to the processor until after all other threads that have the same (or higher) priority have been processed.

3.6 Determinism

As noted in Chapter 1, an important feature of real-time embedded systems is the concept of determinism. The traditional definition of this term is based on the assumption that for each system state and each set of inputs, a unique set of outputs and next state of the system can be determined. However, we strengthen the definition of determinism for real-time embedded systems by requiring that the time necessary to process any task be predictable. In particular, we are less concerned with average response time than we are with worst-case response time. For example, we must be able to guarantee the worst-case response time for each system call in order for a real-time embedded system to be deterministic. In other words, simply obtaining the correct answer is not adequate. We must get the right answer within a specified time frame.

Many RTOS vendors claim their systems are deterministic and justify that assertion by publishing tables of minimum, average, and maximum number of clock cycles required for each system call. Thus, for a given application in a deterministic system, it is possible to calculate the timing for a given number of threads, and determine whether real-time performance is actually possible for that application.

3.7 Kernel

A *kernel* is a minimal implementation of an RTOS. It normally consists of at least a scheduler and a context switch handler. Most modern commercial RTOSes are actually kernels, rather than full-blown operating systems.

3.8 RTOS

An RTOS is an operating system that is dedicated to the control of hardware, and must operate within specified time constraints. Most RTOSes are used in embedded systems.

3.9 Context Switch

A *context* is the current execution state of a thread. Typically, it consists of such items as the program counter, registers, and stack pointer. The term *context switch* refers to the saving of one thread's context and restoring a different thread's context so that it can be executed. This normally occurs as a result of preemption, interrupt handling, time-slicing (see below), cooperative round-robin scheduling (see below), or suspension of a thread because it needs an unavailable resource. When a thread's context is restored, then the thread resumes execution at the point where it was stopped. The kernel performs the context switch operation. The actual code required to perform context switches is necessarily processor-specific.

3.10 Time-Slice

The length of time (i.e., number of timer-ticks) for which a thread executes before relinquishing the processor is called its *time-slice*. When a thread's (optional) time-slice expires in ThreadX, all other threads of the same or higher priority levels are given a chance to execute before the time-sliced thread executes again. Time-slicing provides another form of round-robin scheduling. ThreadX provides optional time-slicing on a per-thread basis. The thread's time-slice is assigned during creation and can be modified during execution. If the time-slice is too short, then the scheduler will waste too much processing time performing context switches. However, if the time-slice is too long, then threads might not receive the attention they need.

3.11 Interrupt Handling

An essential requirement of real-time embedded applications is the ability to provide fast responses to asynchronous events, such as hardware or software interrupts. When an interrupt occurs, the context of the executing thread is saved and control is transferred to the appropriate interrupt vector. An *interrupt vector* is an address for an *interrupt service routine (ISR)*, which is user-written software designed to handle or service the needs of a particular interrupt. There may be many ISRs, depending on the number of interrupts that needs to be handled. The actual code required to service interrupts is necessarily processor-specific.

3.12 Thread Starvation

One danger of preemptive, priority-based scheduling is *thread starvation*. This is a situation in which threads that have lower priorities rarely get to execute because the processor spends most of its time on higher-priority threads. One method to alleviate this problem is to make certain that higher-priority threads do not monopolize the processor. Another solution would be to gradually raise the priority of starved threads so that they do get an opportunity to execute.

3.13 Priority Inversion

Undesirable situations can occur when two threads with different priorities share a common resource. *Priority inversion* is one such situation; it arises when a higher-priority thread is suspended because a lower-priority thread has acquired a resource needed by the higher-priority thread. The problem is compounded when the shared resource is not in use while the higher-priority thread is waiting. This phenomenon may cause priority

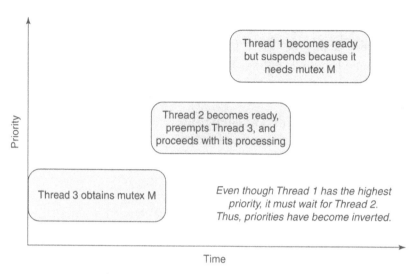

Figure 3.7: Example of priority inversion

inversion time to become nondeterministic and lead to application failure. Consider Figure 3.7, which shows an example of the priority inversion problem.

In this example, Thread 3 (with the lowest priority) becomes ready. It obtains mutex M and begins its execution. Some time later, Thread 2 (which has a higher priority) becomes ready, preempts Thread 3, and begins its execution. Then Thread 1 (which has the highest priority of all) becomes ready. However, it needs mutex M, which is owned by Thread 3, so it is suspended until mutex M becomes available. Thus, the higher-priority thread (i.e., Thread 1) must wait for the lower-priority thread (i.e., Thread 2) before it can continue. During this wait, the resource protected by mutex M is not being used because Thread 3 has been preempted by Thread 2. The concept of priority inversion is discussed more thoroughly in Chapters 8 and 11.

3.14 Priority Inheritance

Priority inheritance is an optional feature that is available with ThreadX for use only with the mutex services. (Mutexes are discussed in more detail in the next chapter.) Priority inheritance allows a lower-priority thread to temporarily assume the priority of a higher-priority thread that is waiting for a mutex owned by the lower-priority thread.

Priority	Comment
0 : 14	Preemption allowed for threads with priorities from 0 to 14 (inclusive)
15 : 19	Thread is assigned preemption-threshold = 15 [this has the effect of disabling preemption for threads with priority values from 15 to 19 (inclusive)]
20 : 31	Thread is assigned Priority = 20

Figure 3.8: Example of preemption-threshold

This capability helps the application to avoid nondeterministic priority inversion by eliminating preemption of intermediate thread priorities. This concept is discussed more thoroughly in Chapters 7 and 8.

3.15 Preemption-Threshold

Preemption-threshold[3] is a feature that is unique to ThreadX. When a thread is created, the developer has the option of specifying a priority ceiling for disabling preemption. This means that threads with priorities greater than the specified ceiling are still allowed to preempt, but those with priorities equal to or less than the ceiling are not allowed to preempt that thread. The preemption-threshold value may be modified at any time during thread execution. Consider Figure 3.8, which illustrates the impact of preemption-threshold. In this example, a thread is created and is assigned a priority value of 20 and a preemption-threshold of 15. Thus, only threads with priorities higher than 15 (i.e., 0 through 14) will be permitted to preempt this thread. Even though priorities 15 through 19 are higher than the thread's priority of 20, threads with those priorities will not be allowed to preempt this thread. This concept is discussed more thoroughly in Chapters 7 and 8.

---

[3]Preemption-threshold is a trademark of Express Logic, Inc. There are several university research papers that analyze the use of preemption-threshold in real-time scheduling algorithms. A complete list of URLs for these papers can be found at http://www.expresslogic.com/news/detail/?prid = 13.

3.16 Key Terms and Phrases

asynchronous event	ready thread
context switch	Ready Thread List
cooperative multithreading	round-robin scheduling
determinism	RTOS
interrupt handling	scheduling
kernel	sleep mode
preemption	suspended thread
preemption-threshold	Suspended Thread List
priority	thread starvation
priority inheritance	time-slice
priority inversion	timer-tick

3.17 Problems

1. When a thread is removed from the Suspended Thread List, either it is placed on the Ready Thread List or it is terminated. Explain why there is not an option for that thread to become the currently executing thread immediately after leaving the Suspended Thread List.

2. Suppose every thread is assigned the same priority. What impact would this have on the scheduling of threads? What impact would there be if every thread had the same priority and was assigned the same duration time-slice?

3. Explain how it might be possible for a preempted thread to preempt its preemptor? Hint: Think about priority inheritance.

4. Discuss the impact of assigning every thread a preemption-threshold value of 0 (the highest priority).

CHAPTER 4

RTOS Building Blocks for System Development

4.1 Introduction

An RTOS must provide a variety of services to the developer of real-time embedded systems. These services allow the developer to create, manipulate, and manage system resources and entities in order to facilitate application development. The major goal of this chapter is to review the services and components that are available with ThreadX. Figure 4.1 contains a summary of these services and components.

4.2 Defining Public Resources

Some of the components discussed are indicated as being *public* resources. If a component is a public resource, it means that it can be accessed from any thread. Note that accessing a component is not the same as owning it. For example, a mutex can be accessed from any thread, but it can be owned by only one thread at a time.

Threads	Message queues	Counting semaphores
Mutexes	Event flags	Memory block pools
Memory byte pools	Application timers	Time counter and interrupt control

Figure 4.1: ThreadX components

4.3 ThreadX Data Types

ThreadX uses special primitive data types that map directly to data types of the underlying C compiler. This is done to ensure portability between different C compilers. Figure 4.2 contains a summary of ThreadX service call data types and their associated meanings.

In addition to the primitive data types, ThreadX uses system data types to define and declare system resources, such as threads and mutexes. Figure 4.3 contains a summary of these data types.

4.4 Thread

A thread is a semi-independent program segment. Threads within a process share the same memory space, but each thread must have its own stack. Threads are the essential building blocks because they contain most of the application programming logic. There is

Data type	Description
UINT	Basic unsigned integer. This type must support 8-bit unsigned data; however, it is mapped to the most convenient unsigned data type, which may support 16- or 32-bit signed data.
ULONG	Unsigned long type. This type must support 32-bit unsigned data.
VOID	Almost always equivalent to the compiler's void type.
CHAR	Most often a standard 8-bit character type.

Figure 4.2: ThreadX primitive data types

System data type	System resource
TX_TIMER	Application timer
TX_QUEUE	Message queue
TX_THREAD	Application thread
TX_SEMAPHORE	Counting semaphore
TX_EVENT_FLAGS_GROUP	Event flags group
TX_BLOCK_POOL	Memory block pool
TX_BYTE_POOL	Memory byte pool
TX_MUTEX	Mutex

Figure 4.3: ThreadX system data types

no explicit limit on how many threads can be created and each thread can have a different stack size. When threads are executed, they are processed independently of each other.

When a thread is created, several attributes need to be specified, as indicated in Figure 4.4. Every thread must have a Thread Control Block (TCB) that contains system information critical to the internal processing of that thread. However, most applications have no need to access the contents of the TCB. Every thread is assigned a name, which is used primarily for identification purposes. The thread entry function is where the actual C code for a thread is located. The thread entry input is a value that is passed to the thread entry function when it first executes. The use for the thread entry input value is determined exclusively by the developer. Every thread must have a stack, so a pointer to the actual stack location is specified, as well as the stack size. The thread priority must be specified but it can be changed during run-time. The preemption-threshold is an optional value; a value equal to the priority disables the preemption-threshold feature. An optional time-slice may be assigned, which specifies the number of timer-ticks that this thread is allowed to execute before other ready threads with the same priority are permitted to run. Note that use of preemption-threshold disables the time-slice option. A time-slice value of zero (0) disables time-slicing for this thread. Finally, a start option must be specified that indicates whether the thread starts immediately or whether it is placed in a suspended state where it must wait for another thread to activate it.

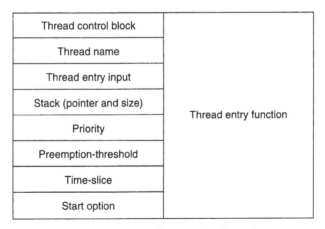

Figure 4.4: Attributes of a thread

4.5 Memory Pools

Several resources require allocation of memory space when those resources are created. For example, when a thread is created, memory space for its stack must be provided. ThreadX provides two memory management techniques. The developer may choose either one of these techniques for memory allocation, or any other method for allocating memory space.

The first of the memory management techniques is the memory byte pool, which is illustrated in Figure 4.5. As its name implies, the memory byte pool is a sequential collection of bytes that may be used for any of the resources. A memory byte pool is similar to a standard C heap. Unlike the C heap, there is no limit on the number of memory byte pools. In addition, threads can suspend on a pool until the requested memory is available. Allocations from a memory byte pool are based on a specified number of bytes. ThreadX allocates from the byte pool in a *first-fit* manner, i.e., the first free memory block that satisfies the request is used. Excess memory from this block is converted into a new block and placed back in the free memory list, often resulting in *fragmentation*. ThreadX *merges* adjacent free memory blocks together during a subsequent allocation search for a large enough block of free memory. This process is called *defragmentation*.

Figure 4.6 contains the attributes of a memory byte pool. Every memory byte pool must have a Control Block that contains essential system information. Every memory byte

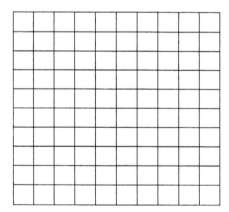

Figure 4.5: Memory byte pool

pool is assigned a name, which is used primarily for identification purposes. The starting address of the byte pool must be provided, as well as the total number of bytes to be allocated to the memory byte pool.

The second type of memory management technique is the memory block pool, which is illustrated in Figure 4.7. A memory block pool consists of fixed-size memory blocks, so there is never a fragmentation problem. There is a lack of flexibility because the same amount of memory is allocated each time. However, there is no limit as to how many memory block pools can be created, and each pool could have a different memory block size. In general, memory block pools are preferred over memory byte pools because the fragmentation problem is eliminated and because access to the pool is faster.

Figure 4.8 contains the attributes of a memory block pool. Every memory block pool must have a Control Block that contains important system information. Every memory block pool is assigned a name, which is used primarily for identification purposes. The number of bytes in each fixed-size memory block must be specified. The address where the memory block pool is located must be provided. Finally, the total number of bytes available to the entire memory block pool must be indicated.

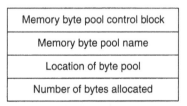

Figure 4.6: Attributes of a memory byte pool

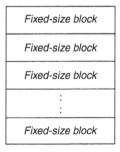

Figure 4.7: Memory block pool

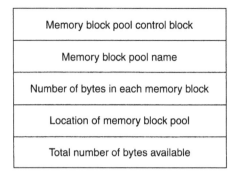

Memory block pool control block
Memory block pool name
Number of bytes in each memory block
Location of memory block pool
Total number of bytes available

Figure 4.8: Attributes of a memory block pool

The total number of memory blocks in a memory block pool can be calculated as follows:

$$Total\ Number\ of\ Blocks = \frac{Total\ Number\ of\ Bytes\ Available}{(Number\ of\ Bytes\ in\ Each\ Memory\ Block) + (sizeof(void*))}$$

Each memory block contains one pointer of overhead that is invisible to the user and is represented by the *sizeof (void*)* expression in the preceding formula. Avoid wasting memory space by correctly computing the total number of bytes to allocate, based on the number of desired memory blocks.

4.6 Application Timer

Fast response to asynchronous external events is the most important function of real-time, embedded applications. However, many of these applications must also perform certain activities at predetermined intervals of time. Application timers enable applications to execute application C functions at specific intervals of time. It is also possible for an application timer to expire only once. This type of timer is called a *one-shot timer,* while repeating interval timers are called *periodic timers*. Each application timer is a public resource.

Figure 4.9 contains the attributes of an application timer. Every application timer must have a Control Block that contains essential system information. Every application timer is assigned a name, which is used primarily for identification purposes. Other attributes include the name of the expiration function that is executed when the timer expires. Another attribute is a value that is passed to the expiration function. (This value is for

| Application timer control block |
| Application timer name |
| Expiration function to call
Expiration input value to pass to function |
| Initial number of timer-ticks |
| Reschedule number of timer-ticks |
| Automatic activate option |

Figure 4.9: Attributes of an application timer

the use of the developer.) An attribute containing the initial number of timer-ticks[1] for the timer expiration is required, as is an attribute specifying the number of timer-ticks for all timer expirations after the first. The last attribute is used to specify whether the application timer is automatically activated at creation, or whether it is created in a nonactive state that would require a thread to start it.

Application timers are very similar to ISRs, except the actual hardware implementation (usually a single periodic hardware interrupt is used) is hidden from the application. Such timers are used by applications to perform time-outs, periodic operations, and/ or watchdog services. Just like ISRs, application timers most often interrupt thread execution. Unlike ISRs, however, application timers cannot interrupt each other.

4.7 Mutex

The sole purpose of a mutex is to provide mutual exclusion; the name of this concept provides the derivation of the name mutex (i.e., MUTual EXclusion).[2] A mutex is used

---

[1]The actual time between timer-ticks is specified by the application, but 10 ms is the value used here.

[2]In the 1960s, Edsger Dijkstra proposed the concept of a mutual exclusion semaphore with two operations: the P operation (Prolaag, meaning to lower) and the V operation (Verhogen, meaning to raise). The P operation decrements the semaphore if its value is greater than zero, and the V operation increments the semaphore value. P and V are atomic operations.

```
┌─────────────────────────────┐
│    Mutex control block       │
├─────────────────────────────┤
│       Mutex name             │
├─────────────────────────────┤
│  Priority inheritance option │
└─────────────────────────────┘
```

Figure 4.10: Attributes of a mutex

to control the access of threads to critical section or certain application resources. A mutex is a public resource that can be owned by one thread only. There is no limit on the number of mutexes that can be defined. Figure 4.10 contains a summary of the attributes of a mutex.

Every mutex must have a Control Block that contains important system information. Every mutex is assigned a name, which is used primarily for identification purposes. The third attribute indicates whether this mutex supports priority inheritance. Priority inheritance allows a lower-priority thread to temporarily assume the priority of a higher-priority thread that is waiting for a mutex owned by the lower-priority thread. This capability helps the application to avoid nondeterministic priority inversion by eliminating preemption of intermediate thread priorities. The mutex is the only ThreadX resource that supports priority inheritance.

4.8 Counting Semaphore

A counting semaphore is a public resource. There is no concept of ownership of semaphores, as is the case with mutexes. The primary purposes of a counting semaphore are event notification, thread synchronization, and mutual exclusion.[3] ThreadX provides 32-bit counting semaphores where the count must be in the range from 0 to 4,294,967,295 or $2^{32}-1$ (inclusive). When a counting semaphore is created, the count must be initialized to a value in that range. Each value in the semaphore is an *instance* of that semaphore. Thus, if the semaphore count is five, then there are five instances of that semaphore.

---

[3]In this instance, mutual exclusion is normally achieved with the use of a binary semaphore, which is a special case of a counting semaphore where the count is restricted to the values zero and one.

Counting semaphore control block
Counting semaphore name
Semaphore count

Figure 4.11: Attributes of a counting semaphore

Event flags group control block
Event flags group name
Group of 32 one-bit event flags

Figure 4.12: Attributes of an event flags group

Figure 4.11 contains the attributes of a counting semaphore. Every counting semaphore must have a Control Block that contains essential system information. Every counting semaphore is assigned a name, which is used primarily for identification purposes. Every counting semaphore must have a Semaphore Count that indicates the number of instances available. As noted above, the value of the count must be in the range from 0x00000000 to 0xFFFFFFFF (inclusive). A counting semaphore can be created either during initialization or during run-time by a thread. There is no limit to the number of counting semaphores that can be created.

4.9 Event Flags Group

An event flags group is a public resource. Event flags provide a powerful tool for thread synchronization. Each event flag is represented by a single bit, and event flags are arranged in groups of 32. When an event flags group is created, all the event flags are initialized to zero.

Figure 4.12 contains the attributes of an event flags group. Every event flags group must have a Control Block that contains essential system information. Every event flags group is assigned a name, which is used primarily for identification purposes. There must also be a group of 32 one-bit event flags, which is located in the Control Block.

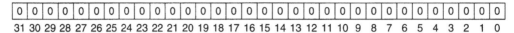

31 30 29 28 27 26 25 24 23 22 21 20 19 18 17 16 15 14 13 12 11 10 9 8 7 6 5 4 3 2 1 0

Figure 4.13: An event flags group

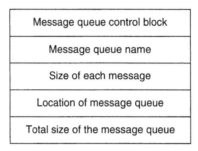

| Message queue control block |
| Message queue name |
| Size of each message |
| Location of message queue |
| Total size of the message queue |

Figure 4.14: Attributes of a message queue

Event flags provide a powerful tool for thread synchronization. Threads can operate on all 32 event flags at the same time. An event flags group can be created either during initialization or during run-time by a thread. Figure 4.13 contains an illustration of an event flags group after it has been initialized. There is no limit to the number of event flags groups that can be created.

4.10 Message Queue

A message queue is a public resource. Message queues are the primary means of interthread communication. One or more messages can reside in a message queue. A message queue that holds a single message is commonly called a *mailbox*. Messages are placed at the rear of the queue,[4] and are removed from the front of the queue.

Figure 4.14 contains the attributes of a message queue. Every message queue must have a Control Block that contains essential system information. Every message queue is assigned a name, which is used primarily for identification purposes. Other attributes include the message size, the address where the message queue is located, and the total number of bytes allocated to the message queue. If the total number of bytes allocated to the message queue is not evenly divisible by the message size, then the remaining bytes are not used.

Figure 4.15 contains an illustration of a message queue. Any thread may insert a message in the queue (if space is available) and any thread may remove a message from a queue.

---

[4]It is also possible to insert a message at the front of the queue.

Messages inserted at rear of queue   Messages removed from front of queue

Figure 4.15: A message queue

	Mutex	Counting semaphore
Speed	Somewhat slower than a semaphore	A semaphore is generally faster than a mutex and requires fewer system resources
Thread ownership	Only one thread can own a mutex	No concept of thread ownership for a semaphore – any thread can decrement a counting semaphore if its current count exceeds zero
Priority inheritance	Available only with a mutex	Feature not available for semaphores
Mutual exclusion	Primary purpose of a mutex – a mutex should be used only for mutual exclusion	Can be accomplished with the use of a binary semaphore, but there may be pitfalls
Inter-thread synchronization	Do not use a mutex for this purpose	Can be performed with a semaphore, but an event flags group should be considered also
Event notification	Do not use a mutex for this purpose	Can be performed with a semaphore
Thread suspension	Thread can suspend if another thread already owns the mutex (depends on value of wait option)	Thread can suspend if the value of a counting semaphore is zero (depends on value of wait option)

Figure 4.16: Comparison of a mutex with a counting semaphore

4.11 Summary of Thread Synchronization and Communication Components

Similarities exist between a mutex and a counting semaphore, especially when implementing mutual exclusion. In particular, a binary semaphore has many of the same properties as that of a mutex. Figure 4.16 contains a comparison of these two resources and recommendations as to how each should be used.

	Thread synchronization	Event notification	Mutual exclusion	Inter-Thread communication
Mutex			Preferred	
Counting semaphore	OK – better for one event	Preferred	OK	
Event flags group	Preferred	OK		
Message queue	OK	OK		Preferred

Figure 4.17: Recommended uses of resources

We discussed four public resources that a thread can use for various purposes, as well as four types of situations where each can be useful. Figure 4.17 contains a summary of the recommended uses of these resources.

4.12 Key Terms and Phrases

application timer	mutex ownership
binary semaphore	mutual exclusion
Control Block	one-shot timer
counting semaphore	periodic timer
defragmentation	preemption
entry function	preemption-threshold
event flags group	primitive data type
event notification	priority
first-fit allocation	priority inheritance
fragmentation	public resource
heap	service call
ISR	stack
mailbox	system data type
memory block pool	thread
memory byte pool	thread suspension
message queue	thread synchronization
mutex	watchdog timer

4.13 Problems

1. Explain why the special primitive data types UINT, ULONG, VOID, and CHAR are used for service calls, rather than the standard C primitive data types.

2. What is the purpose of the thread entry function?

3. Under what circumstances would you use a binary semaphore rather than a mutex for mutual exclusion?

4. There is only one public resource that can be owned by a thread. Which resource is that?

5. Suppose you have a choice in using either a memory byte pool or a memory block pool. Which should you choose? Justify your answer.

6. What does it mean to get an instance of a counting semaphore?

7. What is the maximum number of numeric combinations that can be represented by an event flags group?

8. Messages are usually added to the rear of a message queue. Why would you want to add a message to the front of a message queue?

9. Discuss the differences between a one-shot timer and a periodic timer.

10. What is a timer-tick?

CHAPTER 5

Introduction to the MIPS Microprocessor

5.1 Introduction

The MIPS microprocessor is one of the world's most popular processors for embedded applications. It can be found in applications ranging from hard disk controllers to laser-jet printers to gaming consoles. The simplicity of the MIPS design is an important reason for its success. A simpler processor is easier to use and frequently has faster performance. Because of such features, MIPS is used in many modern products, including the following:

set-top boxes	disk drives
PDAs	medical devices
digital cameras	automobile navigation systems
ink and laser printers	smart cards
switches and routers	modems
wireless devices	game consoles

5.2 History

The MIPS processor ancestry dates back to the pioneering days of RISC (*reduced instruction set computer*) development in the early 1980s by John L. Hennessy of Stanford University. Originally, the MIPS acronym stood for *Microprocessor without Interlocked Pipeline Stages*, but the name is no longer considered an acronym today.[1]

---

[1]Depending on the specific processor, MIPS uses a three-stage, a five-stage, or a six-stage instruction pipeline. Branch instructions flush and refill the pipeline.

In 1984 Hennessy left Stanford and formed MIPS Computer Systems,[2] which focused on the MIPS architecture. The first MIPS processor was called the R2000 and was released in 1985. In 1988, the R3000 was introduced and then the R4000 in 1991. The R4000 was the first 64-bit MIPS microprocessor. The 1990s through the early 2000s also saw the introduction of the R4400, R5000, R8000, R10000, R12000, RM7000, R14000, and R16000.

Today, there are primarily two basic architectures of the MIPS core, the MIPS32 and the MIPS64. The MIPS32 is somewhat similar to the R4000, but has 32-bit registers and addressing, while the MIPS64 has 64-bit registers and addressing. In addition, there are also hyperthreading versions of the MIPS cores, including the MIPS32 34K.

The MIPS microprocessor is a major success. Most people cannot go a day without using a MIPS-based processor.

5.3 Technical Features

As mentioned previously, the MIPS architecture is based on the RISC concept. The driving force behind RISC is that most high-level languages (such as C and C++) can be implemented by using a small set of native processor instructions. When the number and complexity of instructions is small, building the processor is much easier. Furthermore, the processor requires much less power and can execute a simple instruction much faster than a powerful but inherently complex instruction. Following are some basic attributes of the MIPS architecture:

- Load-Store Architecture: Many MIPS instructions operate only on data already stored in registers, which are inherently simple and fast operations. There is a limited number of instructions that move data in memory to and from the registers, thus reducing the complexity of the instruction set.

- Fixed Length Instructions: All MIPS instructions are either 4 bytes or 2 bytes (MIPS16 extension) in length. This eliminates the need to calculate the instruction size and the potential for having to make multiple memory accesses to complete a single instruction fetch.

- Orthogonal Registers: Most MIPS registers can be used for address or data.

---

[2]The company is now known as MIPS Technologies, Inc.

- Single Cycle Execution: Most MIPS instructions execute in a single processor cycle. Obvious exceptions include the load and store instructions mentioned previously.

5.3.1 System-on-Chip (SoC) Compatibility

Miniaturization has been a trend for many years, especially in electronics. There are many reasons for this phenomenon, but important reasons include the drive to reduce the production cost of high-volume products, the need for reduced power consumption, and the pursuit of improved efficiency. Essentially, fewer raw materials translate to lower production cost. In the embedded electronics industry, it has become popular to place many components—processor, memory, and peripherals—on the same chip. This technology is called System-on-Chip (SoC) and it is the primary reason that many devices, such as cell phones, are so much smaller and less expensive than those of the past.

The simplicity of the MIPS architecture makes it a popular processor for SoC designs. Even more important is the MIPS architecture licensing model adopted in the early 1990s. This model is designed and licensed for SoC applications, so major SoC manufacturers commonly have MIPS licenses.

5.3.2 Reduced Power Consumption

Many consumer electronic products are battery powered. Accordingly, the underlying processor and software must be very power efficient. The MIPS architecture is simpler and has fewer registers than most other RISC architectures. Because of these characteristics, it requires less power. Another advantage that most MIPS products have is a feature called *low power mode*. This is a power conservation state initiated by the software when it determines there is nothing important to do. During the low power mode, a small amount of power is used to keep the system coherent. When an interrupt occurs, signaling the arrival of something important, the processor automatically returns to its normal state of operation.

5.3.3 Improved Code Density

One common problem with RISC architectures is low *code density*. Code density is a rough measure of how much work a processor can perform versus program size. Because RISC instructions are simpler than those of *complex instruction set computers*

(CISC), sometimes more RISC instructions are required to perform the same higher-level function. This results in a larger program image, or lower code density.

The 32-bit fixed size instructions of the early MIPS architectures suffered from this problem. A program compiled for execution on a CISC processor could be 30 percent smaller than one compiled for a MIPS architecture (or any RISC processor for that matter). In an attempt to address this problem, MIPS introduced, in the MIPS16 architecture extension, a 16-bit fixed instruction size. The processor recognizes both the fixed-length 16-bit instruction set as well as the original 32-bit MIPS instruction set. A program compiled in MIPS16 is as small as or smaller than the compiled version for a CISC machine.

5.3.4 Versatile Register Set

The MIPS architecture has a total of thirty-two 32-bit general purpose registers, in addition to the following main control registers: one dedicated Program Counter (PC), one dedicated Status Register (Status), one interrupt/exception Cause Register (Cause), one timer Count Register (Count), one timer Compare Register (Compare), and one dedicated Exception Program Counter (EPC). Figure 5.1 contains a description of the MIPS register set.

Register number	Register name	Register usage
$0	$zero	Constant 0
$1	$at	Assembler temporary register
$2–$3	$v0–$v1	Return values
$4–$7	$a0–$a3	Function arguments
$8–$15	$t0–$t7	Temporary registers
$16–$23		Preserved registers
$24–$25	$t8–$t9	Temporary registers
$26–$27	$k0–$k1	Reserved for OS
$28	$gp	Global base pointer
$29	$sp	Stack pointer
$30	$fp	Frame pointer
$31	$ra	Return address

Figure 5.1: MIPS register usage

Each of the general-purpose registers are 32 bits in MIPS32 architectures and 64 bits in MIPS64 architectures. The size of the Program Counter (PC) and Exception Program Counter (EPC) control registers are also determined by the architecture—32 bits in MIPS32 and 64 bits in MIPS64. The main control register in the MIPS architecture is the Status Register (CP0 register 12). Most execution control is determined via bits in the Status Register, including the current mode of execution.

5.3.5 Register Definitions

There are four registers and one counter that are particularly important for ThreadX. They are the Status Register, the Count Register, the Compare Register, the Cause Register, and the Exception Program Counter. Following are descriptions of each.

The Status Register contains all the important information pertaining to the state of execution. Figure 5.2 contains an overview of the MIPS Status Register (CP0 register 12).

Bits	Content
31–28	CU0–CU3
27	RP
26	Reserved
25	RE
24–23	Reserved
22	BEV
21	TS
20	SR
19	NMI
18–16	Reserved
15–8	IM0–IM7
7–5	Reserved
4	UM
3	Reserved
2	ERL
1	EXL
0	IE

Figure 5.2: Status Register (CP0 register 12) overview

Bit position	Name		Meaning
0	IE	Interrupt Enable:	0 → Interrupts Disabled 1 → Interrupts Enabled
1	EXL	Exception Level:	0 → Normal Level 1 → Exception Level
2	ERL	Error Level:	0 → Normal Level 1 → Error Level
3	Reserved	Value of 0	
4	UM	User Mode:	0 → Kernel Mode 1 → User Mode
5–7	Reserved	Value of 0	
8–15	IM0–IM7	Individual Interrupt Enable:	0 → Interrupt Disabled 1 → Interrupt Enabled
16–18	Reserved	Value of 0	
19	NMI	NMI Reset:	0 → Not NMI Reset 1 → NMI Reset
20	SR	Soft Reset:	0 → Not Soft Reset 1 → Soft Reset
21	TS	TLB Shutdown	
22	BEV	Vector Location:	0 → Normal Vector Location 1 → Bootstrap Vector Location
23–24	Reserved	Value of 0	
25	RE	Enable Reverse Endian References	0 → User Mode Uses Default 1 → User Mode Reversed Endianness
26	Reserved	Value of 0	
27	RP	Enables Reduced Power Mode	
28–31	CU0–CU3	Coprocessor Enable:	0 → Coprocessor Access Not Allowed 1 → Coprocessor Access Allowed

Figure 5.3: Status Register (CP0 register 12) definition

Figure 5.3 contains a description of the names, values, and meanings of the fields of the Status Register.

The Count Register is used as a timer and is incremented by one on every other clock. In conjunction with the Compare Register, it is useful in generating a periodic interrupt source. In addition, the Count Register is often read and even written by diagnostic software. It is also important to note that incrementing the Count Register can be disabled while in debug mode by writing to the CountDM bit in the Debug register. Figure 5.4 contains a description of the Count Register.

Fields		Description	Read/ Write	Reset State
Name	Bits			
Count	31:0	Interval counter.	R/W	Undefined

Figure 5.4: Count Register (CP0 register 9) description

Fields		Description	Read/ Write	Reset State
Name	Bit(s)			
Compare	31:0	Interval count compare value.	R/W	Undefined

Figure 5.5: Compare Register (CP0 register 11) description

The Compare Register is used in conjunction with the Count Register to generate timer interrupts. When the Count Register reaches the value in the Compare Register an interrupt is generated. Upon a timer interrupt, it is the responsibility of software to update the Count and/or Compare Register in order to generate another timer interrupt. Figure 5.5 contains a description of the Compare Register.

The Cause Register indicates the source of the most recent interrupt or exception. Upon entering the exception handler, software reads the Cause Register to determine what processing is needed for the particular interrupt or exception. In addition, the Cause Register also contains a limited amount of configuration information, such as the interrupt vector location. Figure 5.6 contains a description of the field names, values, and meanings for the Cause Register.

The Exception Program Counter (EPC) contains the address of the program execution at the beginning of the most recent interrupt or exception. This address is used by the software to return from an interrupt or exception. Figure 5.7 contains a description of the EPC.

5.3.6 Processor Modes

There are four basic processor modes in the MIPS architecture. Some modes are for normal program execution, some are for interrupt processing, and others are for handling program exceptions. The EXL, ERL, and UM of the Status Register define the current processor mode. Figure 5.8 shows these values and their associated processor modes.

Fields		Description
Name	Bits	
BD	31	Indicates whether the last exception taken occurred in a branch delay slot.
TI	30	Timer Interrupt.
CE	29..28	Coprocessor unit number referenced.
DC	27	Disable *Count* Register.
PCI	26	Performance Counter Interrupt.
IV	23	Indicates whether an interrupt exception uses the general exception vector or a special interrupt vector.
WP	22	Indicates that a watch exception was deferred.
IP7..IP2	15..10	Indicates an interrupt is pending.
RIPL	15..10	Requested Interrupt Priority Level.
IP1..IP0	9..8	Controls the request for software interrupts.
ExcCode	6..2	Exception code.
0	25..24, 21..16, 7, 1..0	Must be written as zero; returns zero on read.

Figure 5.6: Cause Register (CP0 register 13) description

Fields		Description	Read/ Write	Reset State
Name	Bit(s)			
EPC	31:0	Exception Program Counter	R/W	Undefined

Figure 5.7: Exception Program Counter (CP0 register 14) description

5.3.6.1 User Program Mode (Status.UM = 1)

This is one of several program execution modes. Because access to system registers is not allowed in this mode, it is typically used by larger operating systems when executing application level programs.

5.3.6.2 Exception Mode (Status.EXL = 1)

This is the mode in which interrupts on the MIPS architecture are processed. The processor stays in this mode until an "eret" instruction is executed or until the bit is manually modified by the software. Note that typical applications have multiple interrupt sources. In such cases, the software—after saving some of the registers on the stack—must determine which interrupt source is responsible for the interrupt and process it.

	Status Register Bits		
Processor Mode	UM	EXL	ERL
User Program Mode	1	0	0
Exception Mode	don't care	1	0
Kernel Mode	0	0	0
Error Mode	don't care	don't care	1

Figure 5.8: Processor modes

5.3.6.3 Kernel Mode (Status.UM = 0)

This is another typical program execution mode. Most embedded systems execute their programs in this mode.

5.3.6.4 Error Mode (Status.ERL = 1)

This program exception mode is used for handling reset, soft reset, and NMI conditions. Unlike the Exception Mode, the return address of the error is saved in the ErrorEPC register instead of the EPC.

5.4 MIPS Power Saving Support

Most MIPS processors have the ability to enter low power mode. In this mode, the processor is sleeping at the instruction used to enter low power mode and will stay in this mode until an interrupt or a debug event occurs. When such an event occurs, the processor completes the low power instruction and prepares for the interrupt just as it would in normal processing.

ThreadX applications typically enter low power mode when the system is idle or when a low priority application thread executes (indicating there is nothing else meaningful to do). The only difficult aspect of entering low power mode is determining if there are any periodic events currently scheduled. ThreadX supports the low power mode processing by providing two utilities, namely `tx_timer_get_next` and `tx_time_ increment`. The `tx_timer_get_next` routine returns the next expiration time. It should be called before entering low power mode and the value returned should be used to reprogram the ThreadX timer (or other timer) to expire at the appropriate time. The `tx_time_increment` utility is used when the processor awakes to adjust the internal

ThreadX timer to the number of timer-ticks that have expired while the processor was in low power mode. By using these two services, the processor can enter low power mode for significant periods of time and without losing any accuracy of ThreadX time-related features.

5.5 Key Terms and Phrases

Cause Register	low power mode
CISC	MIPS 32-bit mode
code density	MIPS 64-bit mode
Compare Register	MIPS architecture
Count Register	orthogonal registers
Error Mode	power saving
exception handler	Program Counter
Exception Mode	register set
Exception Program Counter	RISC
exceptions	single cycle execution
fixed length instructions	SoC
general purpose registers	Status Register
instruction pipeline	System-on-Chip
interrupt handling	timer interrupt
Kernel Mode	User Program Mode
load-store architecture	

CHAPTER 6

MIPS Exception Handling

6.1 Introduction

An *exception* is an asynchronous event or error condition that disrupts the normal flow of thread processing. Usually, an exception must be handled immediately, and then control is returned to thread processing. There are three exception categories in the MIPS architecture, as follows:

- Exceptions resulting from the direct effect of executing an instruction

- Exceptions resulting as a side effect of executing an instruction

- Exceptions resulting from external interrupts, unrelated to instruction execution

When an exception arises, MIPS attempts to complete the current instruction, temporarily halts instruction processing, handles the exception, and then continues to process instructions.

The processor handles an exception by performing the following sequence of actions.

1. Set the EXL bit (bit 1) of the Status CP0 Registers, which disables further interrupts and causes the processor to execute at the Exception Level of execution.

2. Save the current PC (program counter—address of the next instruction) in the EPC register.

3. Change the PC to the appropriate exception vector as illustrated in Figure 6.1, which is where the application software interrupt handling starts.

Address	BEV bit	Vector
0xBFC00000	1	Reset vector—this is where MIPS starts execution on reset or power up.
0xBFC00180	1	This is where MIPS starts executing when an exception or normal interrupt occurs.
0x80000180	0	This is where MIPS starts executing when an exception or normal interrupt occurs.

Figure 6.1: MIPS exception vector area

MIPS has a simple exception and interrupt handling architecture. There are principally two interrupt vectors, one for reset and another for general exception handling. Each vector is an address that corresponds to where the processor starts execution upon the exception condition. Figure 6.1 shows the standard MIPS vector area.

Some implementations of the MIPS architecture add additional interrupt vectors so that each interrupt source can have a separate vector. In addition, some of these separate vectors are given a shadow register set for improved interrupt performance. This scheme has the advantage that the interrupt-handling software no longer has to determine which interrupt source caused the interrupt.

6.2 ThreadX Implementation of MIPS Exception Handling

ThreadX is a popular RTOS for embedded designs using the MIPS processor. ThreadX complements the MIPS processor because both are extremely simple to use and are very powerful.

6.2.1 Reset Vector Initialization

ThreadX initialization on the MIPS processor is straightforward. The reset vector at address 0xBFC00000 contains an instruction that loads the PC with the address of the compiler's initialization routine.[1] Figure 6.2 contains an example of a typical ThreadX vector area, with the reset vector pointing to the entry function _start of the MIPS compiler tools.

---

[1]When you develop an application in C or C++, the compiler generates application initialization code, which must be executed on the target system before the application.

```
        .text
        .globl     _reset_vector              # Address 0xBFC00000
_reset_vector:
        mfc0       $8, $12
        li         $9, 0xFFBFFFFF             # Build mask to clear BEV bit
        and        $8, $8, $9                 # Clear BEV bit
        mtc0       $8, $12                    # Use normal vector area
        lui        $8, %hi(__start)           # Build address of _start routine
        addi       $8, $8, %lo(__start)       #
        j          $8                         # Jump to _start
        nop
        •
        •
        •
        .text
        .globl     _tx_exception_vector
_tx_exception_vector:                         # Address 0x80000180
        la         $26,_tx_exception_handler  # Pickup exception handler address
        j          $26                        # Jump to exception handler
        nop                                   # Delay slot
```

Figure 6.2: ThreadX vector area

There are several different ways to initialize the exception vector. You can set it up by loading it directly to address 0x80000180 with a JTAG debug device. Alternatively, your system may copy the exception vector to address 0x80000180 during initialization.

The exception vector handler is typically located in the ThreadX low-level initialization file tx_initialize_low_level.S and may be modified according to application needs. For example, in many applications, the reset vector will actually point to some low-level application code that is responsible for preparing the hardware memory for execution. This code can also copy itself from flash memory to RAM if that is necessary for the application. Note that ThreadX can execute in place out of flash memory or in RAM. Once finished, the application low-level code must jump to the same location specified in the original reset vector.

For example, suppose a low-level initialization routine called my_low_level_init is required to execute before anything else in the system. The application would have to change the reset vector to point to the routine in Figure 6.3.

```
        .globl   _reset_vector              # Address 0xBFC00000
_reset_vector:
        mfc0    $8, $12                    # Pickup SR
        li      $9, 0xFFBFFFFF             # Build mask to clear BEV bit
        and     $8, $8, $9                 # Clear BEV bit
        mtc0    $8, $12                    # Use normal vector area
        lui     $8, %hi(__my_low_level_init)   # Build address of my init routine
        addi    $8, $8, %lo(__my_low_level_init)
        j       $8                         # Jump to _start
        nop
```

Figure 6.3: Changing the reset vector

```
        .globl   my_low_level_init
_my_low_level_init
                                           # Build address of _start
        lui     $8, %hi(__start)           routine
        addi    $8, $8, %lo(__start)       #
        j       $8                         # Jump to _start
        nop
```

Figure 6.4: ThreadX low-level initialization

At the end of my_low_level_init, the code would have to branch (jump) to call the original compiler startup code, as illustrated in Figure 6.4.

6.2.1.1 Compiler Initialization

Shortly after the MIPS processor executes the reset vector, the system executes the C compiler's initialization code. In this example, the name of the compiler's entry point is __start. The C compiler initialization is responsible for setting up all application data areas, including initialized and uninitialized global C variables. The C run-time library is also set up here, including the traditional heap memory area. If some of the application code was written in C++, the initialization code instantiates all global C++ objects. Once the run-time environment is completely set up, the code calls the application's "main" entry point.

6.2.1.2 ThreadX Initialization

The ThreadX initialization typically occurs from inside the application's main function. Figure 6.5 shows a typical application main function that uses ThreadX. It is important to

```
                    /* Define main entry point. */
                    void main()
                    {
                      /* Enter the ThreadX kernel. */
                      tx_kernel_enter();
                    }
```

Figure 6.5: Typical application main function that uses ThreadX

note that tx_kernel_enter does not return to main. Hence, any code after tx_kernel_enter will never be executed.

When ThreadX is entered via tx_kernel_enter, it performs a variety of actions in preparation for multithreading on the MIPS processor. The first action is to call ThreadX's internal low-level initialization function _tx_initialize_low_level. This function sets up the system stack pointer, which will be used in interrupt processing. This function also ensures that the exception vector is properly initialized at address 0x80000180. Typically, tx_initialize_low_level also sets up the periodic timer interrupt. When the low-level initialization returns, ThreadX initializes all its system components, which includes creating a system timer thread for the management of ThreadX application timers.

After basic ThreadX initialization is complete, ThreadX calls the application's ThreadX initialization routine, tx_application_define. This is where the application can define its initial ThreadX system objects, including threads, queues, semaphores, mutexes, event flags, timers, and memory pools. After tx_application_define returns, the complete system has been initialized and is ready to go. ThreadX starts scheduling threads by calling its scheduler, _tx_thread_schedule.

6.2.2 Thread Scheduling

ThreadX scheduling occurs within a small loop inside _tx_thread_schedule. ThreadX maintains a pointer that always points to the next thread to schedule. This pointer name is _tx_thread_execute_ptr; this pointer is set to NULL when a thread suspends. If all threads are suspended, it stays NULL until an ISR executes and makes a thread ready. While this pointer is NULL, ThreadX waits in a tight loop until it changes as a result of an interrupt event that results in resuming a thread. While this pointer is not NULL, it points to the TX_THREAD structure associated with the thread to be executed.

Scheduling a thread is straightforward. ThreadX updates several system variables to indicate the thread is executing, recovers the thread's saved context, and transfers control back to the thread.

6.2.2.1 Recovering Thread Context

Recovering a thread's context is straightforward. The thread's context resides on the thread's stack and is available to the scheduler when it schedules the thread. The content of a thread's context depends on how the thread last gave up control of the processor. If the thread made a ThreadX service call that caused it to suspend, or that caused a higher-priority thread to resume, the saved thread's context is small and is called a "solicited" context. Alternatively, if the thread was interrupted and preempted by a higher-priority thread via an ISR, the saved thread's context contains the entire visible register set and is called an "interrupt" context.

A solicited thread context is smaller because of the implementation of the C language for the MIPS architecture. The C language implementation divides the MIPS register set into scratch registers and preserved registers. As the name implies, scratch registers are not preserved across function calls. Conversely, the contents of preserved registers are guaranteed to be the same after a function call returns as they were before the function was called. In the MIPS architecture, registers $16–$23 (s0–s7) and $30 (s8) are considered preserved registers. Because the thread suspension call is itself a C function call, ThreadX can optimize context saving for threads that suspend by calling a ThreadX service. The minimal ThreadX solicited context is illustrated in Figure 6.6.

As Figure 6.6 illustrates, the solicited thread context is extremely small and can reside in 56 bytes of stack space. Thus, saving and recovering the solicited thread context is extremely fast.

An interrupt context is required if the thread was interrupted and the corresponding ISR processing caused a higher-priority thread to be resumed. In this situation, the thread context must include all the visible registers. The interrupt thread context[2] is shown in Figure 6.7.

Figure 6.8 contains the MIPS code fragment that restores the interrupt and solicited thread contexts in a ThreadX MIPS application.[3] The determination of which context to

---

[2]This is also called an interrupt thread context stack frame.

[3]These modifiers utilize the status bits set by the compare and other instructions.

TX_THREAD thread_control_block
{
•
•
•

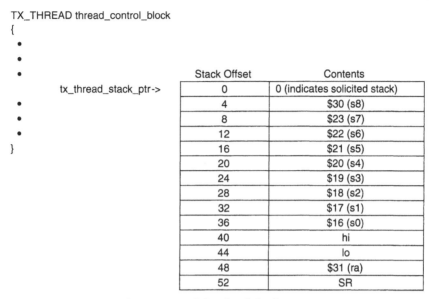

	Stack Offset	Contents
tx_thread_stack_ptr->	0	0 (indicates solicited stack)
	4	$30 (s8)
	8	$23 (s7)
	12	$22 (s6)
	16	$21 (s5)
	20	$20 (s4)
	24	$19 (s3)
	28	$18 (s2)
	32	$17 (s1)
	36	$16 (s0)
	40	hi
	44	lo
	48	$31 (ra)
	52	SR

Figure 6.6: Minimal solicited context

restore is based on the stack type found in the first entry on the stack. Note that $29 is the stack pointer register.

There are two different places in this code example where execution can return to the caller. If the stack type is an interrupt type (contents of the address in $29 has a value of 1), the interrupt context restoration is used. Otherwise, the solicited context recovery code is executed.

The first return method in Figure 6.8 recovers every processor resource for the thread, and the second method recovers only the resources presumed to be saved across function calls. The key point is that what the RTOS must save when a thread makes a function call (ThreadX API call, actually) is much less than what it must save when a thread is interrupted.

6.2.2.2 Saving Thread Context

The saving of a thread's context occurs from within several locations inside the ThreadX RTOS. If an application makes a ThreadX service call that causes a higher-priority thread to preempt the calling thread, the ThreadX service call will call a routine named _tx_thread_system_return to create a solicited thread context on the thread's stack (in the format shown in Figure 6.6) and return to the ThreadX scheduler. Alternatively, if ThreadX detects that a higher-priority thread became ready during the application's ISR

TX_THREAD thread_control_block

{

    •

    •

    •

    tx_thread_stack_ptr->

    •

    •

    •

}

Stack Offset	Contents
0	1 (indicates interrupt stack)
4	$30 (s8)
8	$23 (s7)
12	$22 (s6)
16	$21 (s5)
20	$20 (s4)
24	$19 (s3)
28	$18 (s2)
32	$17 (s1)
36	$16 (s0)
40	hi
44	lo
48	$25 (t9)
52	$24 (t8)
56	$15 (t7)
60	$14 (t6)
64	$13 (t5)
68	$12 (t4)
72	$11 (t3)
76	$10 (t2)
80	$9 (t1)
84	$8 (t0)
88	$7 (a3)
92	$6 (a2)
96	$5 (a1)
100	$4 (a0)
104	$3 (v1)
108	$2 (v0)
112	$1 (at)
116	$31 (ra)
120	SR
124	EPC

Figure 6.7: Interrupt thread context

```
lw          $10, ($29)                          # Pickup stack type
nop                                             # Delay slot
beqz        $10,                                # If 0, solicited thread
            _tx_thread_synch_return             return
nop                                             # Delay slot


/* Recover interrupt context.  */


lw          $8,124($29)                         # Recover EPC
lw          $9,120($29)                         # Recover SR
mtc0        $8, $14                             # Setup EPC
lw          $30, 4($29)                         # Recover s8
mtc0        $9, $12                             # Restore SR
lw          $23, 8($29)                         # Recover s7
lw          $22, 12($29)                        # Recover s6
lw          $21, 16($29)                        # Recover s5
lw          $20, 20($29)                        # Recover s4
lw          $19, 24($29)                        # Recover s3
lw          $18, 28($29)                        # Recover s2
lw          $17, 32($29)                        # Recover s1
lw          $16, 36($29)                        # Recover s0
lw          $8, 40($29)                         # Recover hi
lw          $9, 44($29)                         # Recover low
mthi        $8                                  # Setup hi
mtlo        $9                                  # Setup lo
lw          $25, 48($29)                        # Recover t9
lw          $24, 52($29)                        # Recover t8
lw          $15, 56($29)                        # Recover t7
lw          $14, 60($29)                        # Recover t6
lw          $13, 64($29)                        # Recover t5
lw          $12, 68($29)                        # Recover t4
lw          $11, 72($29)                        # Recover t3
lw          $10, 76($29)                        # Recover t2
lw          $9, 80($29)                         # Recover t1
lw          $8, 84($29)                         # Recover t0
lw          $7, 88($29)                         # Recover a3
lw          $6, 92($29)                         # Recover a2
lw          $5, 96($29)                         # Recover a1
```

Figure 6.8: MIPS code fragment to restore interrupt and solicited thread context

```
        lw       $4, 100($29)              # Recover a0
        lw       $3, 104($29)              # Recover v1
        lw       $2, 108($29)              # Recover v0
        .set     noat
        lw       $1, 112($29)              # Recover at
        .set     at
        lw       $31,116($29)              # Recover ra
        addu     $29, $29, 128             # Recover stack frame

                                           # Return to point of
        eret                               interrupt
        nop                                # Delay
_tx_thread_synch_return:

        /* Recover solicited context. */

        lw       $30, 4($29)               # Recover s8
        lw       $23, 8($29)               # Recover s7
        lw       $22, 12($29)              # Recover s6
        lw       $21, 16($29)              # Recover s5
        lw       $20, 20($29)              # Recover s4
        lw       $19, 24($29)              # Recover s3
        lw       $18, 28($29)              # Recover s2
        lw       $17, 32($29)              # Recover s1
        lw       $16, 36($29)              # Recover s0
        lw       $8,  40($29)              # Recover hi
        lw       $9,  44($29)              # Recover low
        mthi     $8                        # Setup hi
        mtlo     $9                        # Setup lo
        lw       $8,  52($29)              # Recover SR
        lw       $31, 48($29)              # Recover ra
        addu     $29, $29, 56              # Recover stack space
        mtc0     $8, $12                   # Restore SR
        j        $31                       # Return to thread
        nop                                # Delay slot
```

Figure 6.8: Continued

(by application calls to ThreadX services), ThreadX creates an interrupt thread context on the interrupted thread's stack and returns control to the ThreadX scheduler.

Note that ThreadX also creates an interrupt thread context when each thread is created. When ThreadX schedules the thread for the first time, the interrupt thread context contains an interrupt return address that points to the first instruction of the thread.

6.2.3 ThreadX Interrupt Handling

ThreadX provides basic handling for all MIPS program exceptions and interrupts. The ThreadX program exception handlers are small spin loops that enable the developer to easily set a breakpoint and detect immediately when a program exception occurs. These small handlers are located in the low-level initialization code in the file tx_initialize_low_level.S. ThreadX offers full management of all MIPS interrupts, which are described in the following sections.

6.2.3.1 Exception Handling

ThreadX provides full management of MIPS's exceptions and interrupts. As described before, the exception vector starts at address 0x80000180, which typically contains the instructions:

```
        .text
        .globl    _tx_exception_vector

 _tx_exception_vector:                          # Address 0x80000180

        la      $26,_tx_exception_handler      # Pickup exception handler address
        j       $26
                                                # Jump to exception handler
        nop
                                                # Delay slot
```

These instructions jump to _tx_exception_handler, the ThreadX MIPS exception handler. Figure 6.9 contains an example of a basic ThreadX MIPS exception handler.

After _tx_thread_context_save returns, execution is still in the exception mode. The SR, point of interrupt, and all C scratch registers are available for use. At this point, interrupts are still disabled. As illustrated in Figure 6.8, the application's interrupt handlers are

```
        .globl      _tx_exception_handler
_tx_exception_handler:
        mfc0        $26, $13                        # Pickup the cause register
        nop                                         # Delay slot
        andi        $26, $26, 0x3C                  # Isolate the exception code
        bne         $26, $0,_tx_error_exceptions    # If non-zero, an error exception is present
        nop                                         # Delay slot
        /* Otherwise, an interrupt exception is present.  Call context save before
           we process normal interrupts. */

        la          $26, _tx_thread_context_save    # Pickup address of context save function
        jalr        $27, $26                        # Call context save
        nop                                         # Delay slot

        /* Perform interrupt processing here!  When context save returns, interrupts
           are disabled and all compiler scratch registers are available.  Also, s0
           is saved and is used in this function to hold the contents of the CAUSE
           register. */

        mfc0        $16, $13                        # Pickup the cause register
        nop                                         # Delay slot

        /* Interrupts may be re-enabled after this point. */

        /* Check for Interrupt 0. */

        andi        $8, $16, INTERRUPT_0            # Isolate interrupt 0 flag
        beqz        $8, _tx_not_interrupt_0         # If not set, skip interrupt 0 processing
                                                    # Delay slot

        /* Interrupt 0 processing goes here! */

_tx_not_interrupt_0:

        /* Check for Interrupt 1. */

        andi        $8, $16, INTERRUPT_1            # Isolate interrupt 1 flag
        beqz        $8, _tx_not_interrupt_1         # If not set, skip interrupt 1 processing
        nop                                         # Delay slot

        /* Interrupt 1 processing goes here! */

_tx_not_interrupt_1:

        /* Check for Interrupt 2. */

        andi        $8, $16, INTERRUPT_2            # Isolate interrupt 2 flag
        beqz        $8, _tx_not_interrupt_2         # If not set, skip interrupt 2 processing
        nop                                         # Delay slot
```

Figure 6.9: Example of a ThreadX MIPS exception handler

```
        /* Interrupt 2 processing goes here! */
_tx_not_interrupt_2:

        /* Check for Interrupt 3.  */

        andi    $8, $16, INTERRUPT_3            # Isolate interrupt 3 flag
        beqz    $8, _tx_not_interrupt_3         # If not set, skip interrupt 3 processing
        nop                                     # Delay slot

        /* Interrupt 3 processing goes here!  */

_tx_not_interrupt_3:

        /* Check for Interrupt 4.  */

        andi    $8, $16, INTERRUPT_4            # Isolate interrupt 4 flag
        beqz    $8, _tx_not_interrupt_4         # If not set, skip interrupt 4 processing
        nop                                     # Delay slot

        /* Interrupt 4 processing goes here!  */

_tx_not_interrupt_4:

        /* Check for Interrupt 5.  */

        andi    $8, $16, INTERRUPT_5            # Isolate interrupt 5 flag
        beqz    $8, _tx_not_interrupt_5         # If not set, skip interrupt 5 processing
        nop                                     # Delay slot

        /* Interrupt 5 processing goes here!  */

_tx_not_interrupt_5:

        /* Check for Software Interrupt 0.  */

        andi    $8, $16, SW_INTERRUPT_0         # Isolate software interrupt 0 flag
        beqz    $8, _tx_not_interrupt_sw_0      # If not set, skip sw interrupt 0 processing
        nop                                     # Delay slot

        /* Software interrupt 0 processing goes here!  */

_tx_not_interrupt_sw_0:

        /* Check for Software Interrupt 1.  */

        andi    $8, $16, SW_INTERRUPT_1         # Isolate software interrupt 1 flag
        beqz    $8, _tx_not_interrupt_sw_1      # If not set, skip sw interrupt 1 processing
        nop                                     # Delay slot


        /* Software interrupt 1 processing goes here!  */

_tx_not_interrupt_sw_1:


                                                # Pickup address of context restore
        la      $8, _tx_thread_context_restore  function
        i       $8                              # Jump to context restore - does not return!
```

```
_tx_error_exceptions:
        b       _tx_error_exceptions    # Default error exception processing
        nop                             # Delay slot
```

Figure 6.10: Example of a ThreadX error exception handler

called between the ThreadX context save and context restore calls. Of course the interrupt calls must be made from the proper location of the assembly language dispatch routine.

6.2.3.2 Error Exception Handling

ThreadX also provides management of the MIPS's error exceptions. The default handler is simply a spin loop and may be modified to whatever processing is appropriate for the application. Figure 6.10 contains an example of a basic ThreadX Error Exception handler.

Note that ThreadX interrupt management has not yet been called upon entry to the error exception handler. Hence, all registers used in the error exception handler must be saved/restored by the handler. Alternatively, the ThreadX context save/restore routines may also be utilized.

6.2.4 Internal Interrupt Processing

ThreadX interrupt processing is tailored to the MIPS architecture. There are several optimizations and additional interrupt handling features in ThreadX that are not found in other commercial RTOSes. We discuss some of these features below as we describe how ThreadX processes interrupts.

6.2.4.1 Idle System

Unlike other RTOSes that require a background thread to be continuously running, ThreadX implements its idle loop as a simple three-instruction sequence in assembly code. These three instructions are designed to wait for the next thread to be ready for scheduling. There are several advantages to this approach, including not wasting the memory resources associated with having an idle thread—including the thread's Control Block, stack, and instruction area. Note that all the threads in the system still require resources. However, with this idle loop approach, ThreadX need not force the application to maintain a dummy thread that executes when the system is idle. A dummy thread would require a TCB and a stack, and would eliminate an optimization in the interrupt handling because the thread's context would always need to be saved. The other advantage involves interrupt processing. If ThreadX detects that an interrupt has occurred when the system is idle (in the scheduling loop), no context (registers) need to be saved or restored. When the interrupt processing is complete, a simple restart of the scheduling loop will suffice.

6.2.4.2 *Saving Solicited Thread Contexts*

If an interrupt does occur during a thread's execution, ThreadX initially saves only the thread's scratch registers via the _tx_thread_context_save routine. Assuming the ISR contains C calls, the compiler will preserve the nonscratch registers during the ISR processing. If the ISR processing does not make a higher-priority thread ready, the minimal context saved ($1 (at) through $15 (t7), $24 (t8), $25(t9), hi, and lo) is recovered, followed by an interrupt return to the interrupted thread.

6.2.4.3 *Saving Interrupt Thread Contexts*

If an application interrupt service routine makes a higher-priority thread ready, ThreadX builds an interrupt thread context stack frame (see Figure 6.7) and returns to the thread scheduler.

6.2.4.4 *Nested Interrupt Handling*

ThreadX supports nested interrupts on the MIPS architecture. Simple nesting on top of any MIPS interrupts is inherently supported. The application simply must clear the interrupt source and re-enable interrupts in its ISR handling.

6.3 Key Terms and Phrases

MIPS exception handling	nested interrupts
asynchronous event	preserved registers
disable interrupts	recovering thread context
error condition	reset vector
exception	saving thread context
exception categories	scratch registers
interrupt handling	solicited stack frame
idle system	solicited thread context
initialization routine	spin loops
interrupt service routine	thread preemption
interrupt stack frame	ThreadX interrupt handling
interrupt thread context	ThreadX vector area
interrupt vectors	vector area

CHAPTER 7

The Thread—The Essential Component

7.1 Introduction

You have investigated several aspects of threads in previous chapters, including their purpose, creation, composition, and usage. In this chapter, you will explore all the services that directly affect threads. To get started, you will review the purpose as well as the contents of the Thread Control Block. You will also examine each of the thread services, with an emphasis on the features and capabilities of each service.

7.2 Thread Control Block

The Thread Control Block (TCB)[1] is a structure used to maintain the state of a thread during run-time. It is also used to preserve information about a thread when a context switch becomes necessary. Figure 7.1 contains many of the fields that comprise the TCB.

A TCB can be located anywhere in memory, but it is most common to make the Control Block a global structure by defining it outside the scope of any function.[2] Locating the Control Block in other areas requires a bit more care, as is the case for all dynamically allocated memory. If a Control Block were allocated within a C function, the memory

---

[1]The characteristics of each thread are contained in its TCB. This structure is defined in the *tx_api.h* file.

[2]Comments about the storage and use of the TCB are also applicable to Control Blocks for other ThreadX entities.

Field	Description	Field	Description
tx_thread_id	Control block ID	tx_thread_state	Thread's execution state
tx_thread_run_count	Thread's run counter	tx_thread_delayed_suspend	Delayed suspend flag
tx_thread_stack_ptr	Thread's stack pointer	tx_thread_suspending	Thread suspending flag
tx_thread_stack_start	Stack starting address	tx_thread_preempt_threshold	Preemption-threshold
tx_thread_stack_end	Stack ending address	tx_thread_priority_bit	Priority ID bit
tx_thread_stack_size	Stack size	*tx_thread_thread_entry	Thread function entry point
tx_thread_time_slice	Current time-slice	tx_thread_entry_parameter	Thread function parameter
tx_thread_new_time_slice	New time-slice	tx_thread_thread_timer	Thread timer block
*tx_thread_ready_next	Pointer to the next ready thread	*tx_thread_suspend_cleanup	Thread's cleanup function and associated data
*tx_thread_ready_previous	Pointer to the previous ready thread	*tx_thread_created_next	Pointer to the next thread in the created list
tx_thread_thread_name	Pointer to thread's name	*tx_thread_created_previous	Pointer to the previous thread in the created list
tx_thread_priority	Priority of thread (default: 0–31)		

Figure 7.1: Thread Control Block

associated with it would be allocated on the calling thread's stack. In general, avoid using local storage for Control Blocks because once the function returns, then its entire local variable stack space is released—regardless of whether another thread is using it for a Control Block.

tx_thread_run_count	This member contains a count of how many times the thread has been scheduled. An increasing counter indicates the thread is being scheduled and executed.
tx_thread_state	This member contains the state of the associated thread. The following list represents the possible thread states:

TX_READY	0×00
TX_COMPLETED	0×01
TX_TERMINATED	0×02
TX_SUSPENDED	0×03
TX_SLEEP	0×04
TX_QUEUE_SUSP	0×05
TX_SEMAPHORE_SUSP	0×06
TX_EVENT_FLAG	0×07
TX_BLOCK_MEMORY	0×08
TX_BYTE_MEMORY	0×09
TX_MUTEX_SUSP	0×0D

Figure 7.2: Two useful members of the Thread Control Block

In most cases, the developer need not know the contents of the TCB. However, in some situations, especially during debugging, inspecting certain fields (or members) becomes quite useful. Figure 7.2 contains detailed information about two of the more useful TCB fields for developers.

There are many other useful fields in the TCB, including the stack pointer, time-slice value, and priority. The developer may inspect the members of the TCB, but is strictly prohibited from modifying them. There is no explicit value that indicates whether the thread is currently executing. Only one thread executes at a given time, and ThreadX keeps track of the currently executing thread elsewhere. Note that the value of tx_state for an executing thread is TX_READY.

7.3 Summary of Thread Services

Appendices A through J comprise a ThreadX User Guide. Each of these 10 appendices is devoted to a particular ThreadX service. Appendix H contains detailed information about thread services, including the following items for each service: prototype, brief description of the service, parameters, return values, notes and warnings, allowable invocation, preemption possibility, and an example that illustrates how the service can

Thread service	Description
tx_thread_create	Create an application thread
tx_thread_delete	Delete an application thread
tx_thread_entry_exit_notify	Notify application upon thread entry and exit
tx_thread_identify	Retrieves pointer to currently executing thread
tx_thread_info_get	Retrieve information about a thread
tx_thread_performance_info_get	Get thread performance information
tx_thread_performance_system_info_get	Get thread system performance information
tx_thread_preemption_change	Change preemption-threshold of application thread
tx_thread_priority_change	Change priority of an application thread
tx_thread_relinquish	Relinquish control to other application threads
tx_thread_reset	Reset thread to original status
tx_thread_resume	Resume suspended application thread
tx_thread_sleep	Suspend current thread for specified time
tx_thread_stack_error_notify	Register thread stack error notification callback
tx_thread_suspend	Suspend an application thread
tx_thread_terminate	Terminates an application thread
tx_thread_time_slice_change	Changes time-slice of application thread
tx_thread_wait_abort	Abort suspension of specified thread

Figure 7.3: Thread services

be used. Figure 7.3 contains a listing of all available thread services. In the following sections of this chapter, we will study each of these services. We will consider the many features of the services, and we will develop several illustrative examples.

7.4 Thread Creation

A thread is declared with the TX_THREAD data type[3] and is defined with the tx_thread_create service. Each thread must have its own stack; the developer determines the stack size and the manner in which memory is allocated for the stack. Figure 7.4 illustrates a typical thread stack. There are several methods of allocating memory for the stack, including use of byte pools, block pools, and arrays; or simply specifying a physical starting address in

---

[3]When a thread is declared, a Thread Control Block is created.

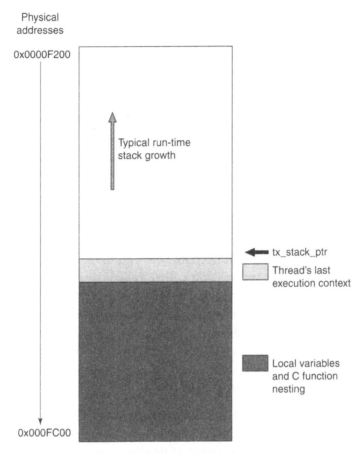

Figure 7.4: Typical thread stack

memory. The stack size is crucial; it must be large enough to accommodate worst-case function call nesting, local variable allocation, and saving the thread's last execution context. The predefined minimum stack size constant, TX_MINIMUM_STACK, is probably too small for most applications. It is better to err toward a larger than a smaller stack.

After a developer has debugged the application, he/she can fine-tune the stack in an attempt to reduce its size. One technique for determining stack space needed is to preset all stack areas with an easily identifiable data pattern, such as 0xEFEF, prior to creating the threads. After thoroughly testing the application, you can deduce how much space was actually used by finding the area of the stack where the preset pattern is still intact.

Threads can require stacks that are quite large. Therefore, it is important to design applications that create a reasonable number of threads and that avoid excessive stack usage within threads. Developers should generally avoid recursive algorithms and large local data structures.

What happens when a stack area is too small? In most cases, the run-time environment simply assumes there is enough stack space. This causes thread execution to corrupt memory adjacent to (usually before) its stack area. The results are very unpredictable, but most often include an unnatural change in the program counter. This is often called *jumping into the weeds*. Of course, the only way to prevent this problem is to ensure that all thread stacks are large enough.

An important feature of multithreading is that the same C function can be called from multiple threads. This feature provides considerable versatility and also helps reduce code space. However, it does require that C functions called from multiple threads be reentrant. A reentrant function is one that can be safely called while it is already being executed. This would happen if, for example, the function were being executed by the current thread and then called again by a preempting thread.[4] To achieve reentrancy, a function stores the caller's return address on the current stack (as opposed to storing it, say, in a register) and does not rely on global or static C variables that it has previously set up. Most compilers do place the return address on the stack. Hence, application developers need only worry about the use of *globals* and *statics*.

An example of a non-reentrant function is the string token function strtok found in the standard C library. This function remembers the previous string pointer on subsequent calls by saving the pointer in a static variable. If this function were called from multiple threads, it would most likely return an invalid pointer.

Chapter 4 illustrates the various building blocks available in ThreadX, including thread attributes. For convenience, the attributes of a thread are illustrated again in Figure 7.5. We will use the tx_thread_create service to create several threads in order to illustrate these attributes.

For the first thread creation example, we will create a thread of priority 15 whose entry point is *"my_thread_entry."* This thread's stack area is 1,000 bytes in size, starting at address 0x400000. We will not use the preemption-threshold feature and we will disable time-slicing. We will place the thread in a ready state as soon as it is created. We also

---

[4]It can also happen if the function is called recursively.

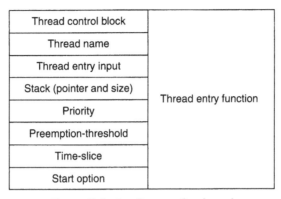

Figure 7.5: Attributes of a thread

```
TX_THREAD my_thread;
UINT status;

/* Create a thread of priority 15 whose entry point is
   "my_thread_entry". This thread's stack area is 1000
   bytes in size, starting at address 0x400000. The
   preemption-threshold is set equal to thread priority to
   disable the preemption threshold feature. Time-slicing
   is disabled. This thread is automatically put into a
   ready condition. */

status = tx_thread_create(&my_thread, "my_thread",
                          my_thread_entry, 0x1234,
                          (VOID *) 0x400000, 1000,
                          15, 15, TX_NO_TIME_SLICE,
                          TX_AUTO_START);

/* If status equals TX_SUCCESS, my_thread is ready
   for execution */

...

/* Thread entry function - When "my_thread" begins
   execution, control is transferred to this function */

VOID my_thread_entry (ULONG initial_input)
{

/* The real work of the thread, including calls to
   other functions should be done here */

}
```

Figure 7.6: Creating a thread with priority 15

need to create a thread entry function to complete the example. Figure 7.6 contains the code necessary to create a thread and its corresponding entry function.

In this figure, the line

```
TX_THREAD my_thread;
```

is used to define a thread called *my_thread*. Recall that TX_THREAD is a data type used to define a TCB. The line

```
UINT status;
```

declares a variable to store the return value from the service call invocation. Each time we invoke a service call, we will check the value of this status variable to determine whether the call was successful. We will use this convention for all invocations to service calls, not just for thread services. The lines beginning with

```
status = tx_thread_create ( … );
```

Parameter	Description
&my_thread	Pointer to a thread control block (defined by TX_THREAD)
"my_thread"	Pointer to the name of the thread—a user-defined name
my_thread_entry	Name of the thread entry function; when the thread begins execution, control is passed to this function
0x1234	A 32-bit value passed to the thread entry function—this value is reserved for the exclusive use of the application
(VOID *) 0x400000	Starting address of the stack's memory area; we used an actual address for the beginning location of the stack, although we have many choices on how to allocate stack space
1000	Number of bytes in the stack memory area
15	Priority—a value in the range from 0 to 31 (inclusive) must be specified
15	Preemption-threshold—a value equal to the priority disables preemption-threshold
TX_NO_TIME_SLICE	Time-slice option—this means we have disabled time-slicing for this thread
TX_AUTO_START	Initial thread status—this means that the thread starts immediately upon creation

Figure 7.7: Thread create parameters used in previous figure

create the thread, where the parameters specify the characteristics of the thread. Figure 7.7 contains descriptions for these parameters.

We need to create a thread entry function for this thread. In this case, the lines

```
VOID my_thread_entry (ULONG initial_input)
{
…
}
```

define that function. As noted earlier, the real work of the thread, including calls to other functions, occurs in this function. The initial_input value is passed to the function and is used exclusively by the application. Many entry functions are in a "do forever" loop and never return, but if the function does return, then the thread is placed in a "completed" state. If a thread is placed in this state, it cannot be executed again.

Consult the appendices to find thorough descriptions of the parameters for all the service calls, as well as the return values that indicate whether a call was successful, and if not, the exact cause of the problem.

For our next thread creation example, we will create a thread of priority 20, also with an entry point of *"my_thread_entry."* This thread's stack area is 1,500 bytes in size, starting at address *&my_stack*. We will use a preemption-threshold value of 14 and we will disable time-slicing. Note that using preemption-threshold automatically disables time-slicing.

```
TX_THREAD my_thread;
UINT status;

/* Create a thread of priority 20 whose entry point is
   "my_thread_entry". This thread's stack area is 1500
   bytes in size, starting at address &my_stack. The
   preemption-threshold is setup to allow preemption at
   priorities above 14. Time-slicing is disabled. This
   thread is automatically put into a ready condition. */
status = tx_thread_create(&my_thread, "my_thread",
                          my_thread_entry, 0x1234,
                          &my_stack, 1500,
                          20, 14, TX_NO_TIME_SLICE,
                          TX_AUTO_START);
/* If status equals TX_SUCCESS, my_thread is ready
   for execution */
```

Figure 7.8: Creating a thread with priority 20 and preemption-threshold 14

```
TX_THREAD my_thread;
UINT status;

/* Create a thread of priority 18 whose entry point is
   "my_thread_entry". This thread's stack area is 1000
   bytes in size, starting at address &my_stack. The
   preemption-threshold feature is disabled. The time-
   slice is 100 timer-ticks. This thread is automatically
   put into a ready condition. */

status = tx_thread_create(&my_thread, "my_thread",
                          my_thread_entry, 0x1234,
                          &my_stack, 1000,
                          18, 18, 100,
                          TX_AUTO_START);

/* If status equals TX_SUCCESS, my_thread is ready
   for execution */
```

Figure 7.9: Creating a thread with priority 18 and no preemption-threshold

A preemption-threshold value of 14 means that this thread can be preempted only by threads with priorities higher than 14, i.e., priorities from 0 to 13 (inclusive). Figure 7.8 contains the code necessary to create this thread.

For our final thread creation example, we will create a thread of priority 18, again with an entry point of *"my_thread_entry"* and a stack starting at *&my_stack*. This thread's stack area is 1,000 bytes in size. We will not use preemption-threshold value but we will use a time-slice value of 100 timer-ticks. Figure 7.9 contains the code necessary to create this thread. Note that time-slicing does result in a small amount of system overhead. It is useful only in cases in which multiple threads share the same priority. If threads have unique priorities, time-slicing should not be used.

There are eight possible return values for thread creation, but only one indicates a successful thread creation. Make certain that you check the return status after every service call.

7.5 Thread Deletion

A thread can be deleted only if it is in a terminated or completed state. Consequently, this service cannot be called from a thread attempting to delete itself. Typically, this service is called by timers or by other threads. Figure 7.10 contains an example showing how thread *my_thread* can be deleted.

```
TX_THREAD my_thread;
UINT status;
...

/* Delete an application thread whose control block is
   "my_thread." Assume that the thread has already been
   created with a call to tx_thread_create. */
status = tx_thread_delete(&my_thread);

/* If status equals TX_SUCCESS, the application thread
   has been deleted. */
```

Figure 7.10: Deletion of thread my_thread

It is the responsibility of the application to manage the memory area used by the deleted thread's stack, which is available after the thread has been deleted. Furthermore, the application must prevent use of a thread after it has been deleted.

7.6 Identify Thread

The tx_thread_identify service returns a pointer to the currently executing thread. If no thread is executing, this service returns a null pointer. Following is an example showing how this service can be used.

```
my_thread_ptr=tx_thread_identify();
```

If this service is called from an ISR, then the return value represents the thread that was running prior to the executing interrupt handler.

7.7 Get Thread Information

All the ThreadX objects have three services that enable you to retrieve vital information about that object. The first such service for threads—the tx_thread_info_get service—retrieves a subset of information from the Thread Control Block. This information provides a "snapshot" at a particular instant in time, i.e., when the service is invoked. The other two services provide summary information that is based on the gathering of run-time performance data. One service—the tx_thread_performance_info_get service—provides an information summary for a particular thread up to the time the service is invoked. By contrast the tx_thread_performance_system_info_get retrieves an

```
TX_THREAD my_thread;
CHAR *name;
UINT state;
ULONG run_count;
UINT priority;
UINT preemption_threshold;
UINT time_slice;
TX_THREAD *next_thread;
TX_THREAD *suspended_thread;
UINT status;

...

/* Retrieve information about the previously created
   thread "my_thread." */
status =  tx_thread_info_get(&my_thread, &name,
                          &state, &run_count,
                          &priority, &preemption_threshold,
                          &time_slice, &next_thread,
                          &suspended_thread);

/* If status equals TX_SUCCESS, the information requested
   is valid. */
```

Figure 7.11: Example showing how to retrieve thread information

information summary for all threads in the system up to the time the service is invoked. These services are useful in analyzing the behavior of the system and determining whether there are potential problem areas. The tx_thread_info_get[5] service obtains information that includes the thread's current execution state, run count, priority, preemption-threshold, time-slice, pointer to the next created thread, and pointer to the next thread in the suspension list. Figure 7.11 shows how this service can be used.

If the variable *status* contains the value TX_SUCCESS, the information was successfully retrieved.

7.8 Preemption-Threshold Change

The preemption-threshold of a thread can be established when it is created or during run-time. The service tx_thread_preemption_change changes the preemption-threshold of an existing thread. The preemption-threshold prevents preemption of a thread by other

---

[5]By default, only the tx_thread_info_get service is enabled. The other two information-gathering services must be enabled in order to use them.

```
UINT my_old_threshold;
UINT status;

/* Disable all preemption of the specified thread. The
   current preemption-threshold is returned in
   "my_old_threshold". Assume that "my_thread" has
   already been created. */
status = tx_thread_preemption_change(&my_thread, 0,
                                     &my_old_threshold);

/* If status equals TX_SUCCESS, the application thread is
   non-preemptable by another thread. Note that ISRs are
   not prevented by preemption disabling. */
```

Figure 7.12: Change preemption-threshold of thread my_thread

```
TX_THREAD my_thread;
UINT my_old_priority;
UINT status;

...

/* Change the thread represented by "my_thread" to priority 0. */
status = tx_thread_priority_change(&my_thread, 0,
                                   &my_old_priority);

/* If status equals TX_SUCCESS, the application thread is
   now at the highest priority level in the system. */
```

Figure 7.13: Change priority of thread my_thread

threads that have priorities equal to or less than the preemption-threshold value. Figure 7.12 shows how the preemption-threshold value can be changed so that preemption by any other thread is prohibited.

In this example, the preemption-threshold value is changed to zero (0). This is the highest possible priority, so this means that no other threads may preempt this thread. However, this does not prevent an interrupt from preempting this thread. If *my_thread* was using time-slicing prior to the invocation of this service, then that feature would be disabled.

7.9 Priority Change

When a thread is created, it must be assigned a priority at that time. However, a thread's priority can be changed at any time by using this service. Figure 7.13 shows how the priority of thread *my_thread* can be changed to zero (0).

When this service is called, the preemption-threshold of the specified thread is automatically set to the new priority. If a new preemption-threshold is desired, the tx_ thread_preemption_change service must be invoked after the priority change service has completed.

7.10 Relinquish Control

A thread may voluntarily relinquish control to another thread by using the tx_thread_ relinquish service. This action is typically taken in order to achieve a form of *round-robin scheduling*. This action is a cooperative call made by the currently executing thread that temporarily relinquishes control of the processor, thus permitting the execution of other threads of the same or higher priority. This technique is sometimes called *cooperative multithreading*. Following is a sample service call that illustrates how a thread can relinquish control to other threads.

```
tx_thread_relinquish();
```

Calling this service gives all other ready threads at the same priority (or higher) a chance to execute before the tx_thread_relinquish caller executes again.

7.11 Resume Thread Execution

When a thread is created with the TX_DONT_START option, it is placed in a suspended state. When a thread is suspended because of a call to tx_thread_suspend, it is also placed in a suspended state. The only way such threads can be resumed is when another thread calls the tx_thread_resume service and removes them from the suspended state. Figure 7.14 illustrates how a thread can be resumed.

```
TX_THREAD my_thread;
UINT status;

    ...

/* Resume the thread represented by "my_thread". */
status = tx_thread_resume(&my_thread);

/* If status equals TX_SUCCESS, the application thread
   is now ready to execute. */
```

Figure 7.14: Example showing the resumption of thread my_thread

7.12 Thread Sleep

On some occasions, a thread needs to be suspended for a specific amount of time. This is achieved with the tx_thread_sleep service, which causes the calling thread to suspend for the specified number of timer-ticks. Following is a sample service call that illustrates how a thread suspends itself for 100 timer-ticks:

```
status = tx_thread_sleep(100);
```

If the variable *status* contains the value TX_SUCCESS, the currently running thread was suspended (or slept) for the prescribed number of timer-ticks.

7.13 Suspend Thread Execution

A specified thread can be suspended by calling the tx_thread_suspend service. A thread can suspend itself, it can suspend another thread, or it can be suspended by another thread. If a thread is suspended in such a manner, then it must be resumed by a call to the tx_thread_resume service. This type of suspension is called *unconditional suspension*. Note that there are other forms of conditional suspension, e.g., in which a thread is suspended because it is waiting for a resource that is not available, or a thread is sleeping for a specific period of time. Following is a sample service call that illustrates how a thread (possibly itself) can suspend a thread called *some_thread*.

```
status = tx_thread_suspend(&some_thread);
```

If the variable status contains the value TX_SUCCESS, the specified thread is unconditionally suspended. If the specified thread is already suspended conditionally, the unconditional suspension is held internally until the prior suspension is lifted. When the prior suspension is lifted, the unconditional suspension of the specified thread is then performed. If the specified thread is already unconditionally suspended, then this service call has no effect.

7.14 Terminate Application Thread

This service terminates the specified application thread, regardless of whether or not that thread is currently suspended. A thread may terminate itself. A terminated thread cannot be executed again. If you need to execute a terminated thread, then you can either reset[6] it or delete it and then create it again. Following is a sample service call that illustrates how a thread (possibly itself) can terminate thread *some_thread*.

```
status = tx_thread_suspend(&some_thread);
```

If the variable *status* contains the value TX_SUCCESS, the specified thread has been terminated.

7.15 Time-Slice Change

The optional time-slice for a thread may be specified when the thread is created, and it may be changed at any time during execution. This service permits a thread to change its own time-slice or that of another thread. Figure 7.15 shows how a time-slice can be changed.

Selecting a time-slice for a thread means that it will not execute more than the specified number of timer-ticks before other threads of the same or higher priorities are given an opportunity to execute. Note that if a preemption-threshold has been specified, then time-slicing for that thread is disabled.

```
TX_THREAD my_thread;
ULONG my_old_time_slice
UINT status;

...

/* Change the time-slice of thread "my_thread" to 20. */

status = tx_thread_time_slice_change(&my_thread, 20,
                                     &my_old_time_slice);

/* If status equals TX_SUCCESS, the thread time-slice has
   been changed to 20 and the previous time-slice is
   stored in "my_old_time_slice." */
```

Figure 7.15: Example showing a time-slice change for thread my_thread

---

[6]The *tx_thread_reset* service can be used to reset a thread to execute at the entry point defined at thread creation. The thread must be in either a completed state or a terminated state in order to be reset.

7.16 Abort Thread Suspension

In some circumstances, a thread may be forced to wait an unacceptably long time (even forever!) for some resource. The Abort Thread Suspension service assists the developer in preventing such an unwanted situation. This service aborts sleep or any wait-related suspension of the specified thread. If the wait is successfully aborted, a TX_WAIT_ ABORTED value is returned from the service that the thread was waiting on. Note that this service does not release explicit suspension that is made by the tx_thread_suspend service. Following is an example that illustrates how this service can be used:

```
status = tx_thread_wait_abort(&some_thread);
```

If the variable *status* contains the value TX_SUCCESS, the sleep or suspension condition of thread some_thread has been aborted, and a return value of TX_WAIT_ABORTED is available to the suspended thread. The previously suspended thread is then free to take whatever action it deems appropriate.

7.17 Thread Notification Services

There are two thread notification services: the tx_thread_entry_exit_notify service and the tx_thread_stack_error_notify service. Each of these services registers a notification callback function. If the tx_thread_entry_exit_notify service has been successfully invoked for a particular thread, then each time that thread is entered or exited, the corresponding callback function is invoked. The processing of this function is the responsibility of the application. In a similar fashion, if the tx_thread_stack_error_notify service has been successfully invoked for a particular thread, then the corresponding callback function is invoked if a stack error occurs. The processing of this error is the responsibility of the application.

7.18 Execution Overview

There are four types of program execution within a ThreadX application: initialization, thread execution, interrupt service routines (ISRs), and application timers. Figure 7.16 shows each type of program execution.

Initialization is the first type of program execution. Initialization includes all program execution between processor reset and the entry point of the *thread scheduling loop*.

After initialization is complete, ThreadX enters its thread scheduling loop. The scheduling loop looks for an application thread that is ready for execution. When a ready thread is found, ThreadX transfers control to it. Once the thread is finished (or another higher-priority thread becomes ready), execution transfers back to the thread scheduling loop in order to find the next-highest-priority ready thread. This process of continually executing and scheduling threads is the most common type of program execution in ThreadX applications.

Interrupts are the cornerstone of real-time systems. Without interrupts, it would be extremely difficult to respond to changes in the external world in a timely manner. What happens when an interrupt occurs? Upon detection of an interrupt, the processor saves key information about the current program execution (usually on the stack), then transfers control to a predefined program area. This predefined program area is commonly called an interrupt service routine. In most cases, interrupts occur during thread execution (or in the thread scheduling loop). However, interrupts may also occur inside an executing ISR or an application timer.

Application timers are very similar to ISRs, except the actual hardware implementation (usually a single periodic hardware interrupt is used) is hidden from the application. Such

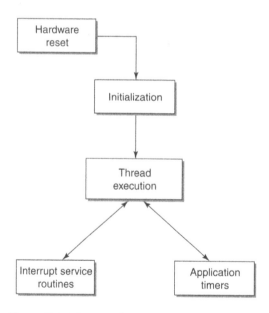

Figure 7.16: Types of program execution

timers are used by applications to perform time-outs, periodic operations, and/or watchdog services. Just like ISRs, application timers most often interrupt thread execution. Unlike ISRs, however, application timers cannot interrupt each other.

7.19 Thread States

Understanding the different processing states of threads is vital to understanding the entire multithreaded environment. There are five distinct thread states: *ready*, *suspended*, *executing*, *terminated*, and *completed*. Figure 7.17 shows the thread state transition diagram for ThreadX.

A thread is in a ready state when it is ready for execution. A ready thread is not executed until it is the highest-priority thread ready. When this happens, ThreadX executes the thread, which changes its state to *executing*. If a higher-priority thread becomes ready, the executing thread reverts back to a *ready* state. The newly ready high-priority thread is then executed, which changes its logical state to *executing*. This transition between *ready* and *executing* states occurs every time thread preemption occurs.

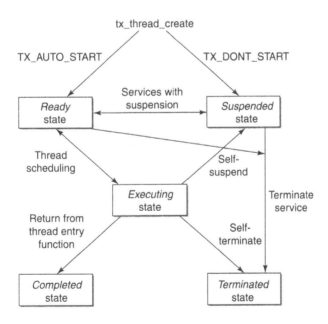

Figure 7.17: Thread state transition

Note that at any given moment only one thread is in an *executing* state. This is because a thread in the *executing* state actually has control of the underlying processor. Threads that are in a *suspended* state are not eligible for execution. Reasons for being in a *suspended* state include suspension for a predetermined time; waiting on message queues, semaphores, mutexes, event flags, or memory; and explicit thread suspension. Once the cause for suspension is removed, the thread returns to a *ready* state.

If a thread is in a *completed* state, this means that the thread has completed its processing and has returned from its entry function. Remember that the entry function is specified during thread creation. A thread that is in a *completed* state cannot execute again unless it is reset to its original state by the tx_thread_reset service.

A thread is in a *terminated* state because another thread called the tx_thread_terminate service, or it called the service itself. A thread in a *terminated* state cannot execute again unless it is reset to its original state by the tx_thread_reset service.

As noted in Chapter 3, threads in the Ready Thread List are eligible for execution. When the scheduler needs to schedule a thread for execution, it selects and removes the thread in that list that has the highest priority and which has been waiting the longest. If an executing thread is interrupted for any reason, its context is saved and it is placed back on the Ready Thread List, ready to resume execution. Threads residing on the Suspended Thread List are not eligible for execution because they are waiting for an unavailable resource, they are in "sleep" mode, they were created with a TX_DONT_START option, or they were explicitly suspended. When a suspended thread has its condition(s) removed, then it is eligible for execution and is moved to the Ready Thread List.

7.20 Thread Design

ThreadX imposes no limits on either the number of threads that can be created or the combinations of priorities that can be used. However, in order to optimize performance and minimize the target size, developers should observe the following guidelines:

- Minimize the number of threads in the application system.
- Choose priorities carefully.
- Minimize the number of priorities.
- Consider preemption-threshold.

- Consider priority inheritance when mutexes are employed.

- Consider round-robin scheduling.

- Consider time-slicing.

There are other guidelines as well, such as making certain that a thread is used to accomplish a particular unit of work, rather than a series of disparate actions.

7.20.1 Minimize the Number of Threads

In general, the number of threads in an application significantly affects the amount of system overhead. This is due to several factors, including the amount of system resources needed to maintain the threads, and the time required for the scheduler to activate the next ready thread. Each thread, whether necessary or not, consumes stack space as well as memory space for the thread itself, and memory for the TCB.

7.20.2 Choose Priorities Carefully

Selecting thread priorities is one of the most important aspects of multithreading. A common mistake is to assign priorities based on a *perceived* notion of thread importance rather than determining what is actually required during run-time. Misuse of thread priorities can starve other threads, create priority inversion, reduce processing bandwidth, and make the application's run-time behavior difficult to understand. If thread starvation is a problem, an application can employ added logic that gradually raises the priority of starved threads until they get a chance to execute. However, properly selecting the priorities in the first place may significantly reduce this problem.

7.20.3 Minimize the Number of Priorities

ThreadX provides 32 distinct priority values that can be assigned to threads. However, developers should assign priorities carefully and should base priorities on the importance of the threads in question. An application that has many different thread priorities inherently requires more system overhead than one with a smaller number of priorities. Recall that ThreadX provides a priority-based, preemptive-scheduling algorithm. This means that lower-priority threads do not execute until there are no higher-priority threads ready for execution. If a higher-priority thread is always ready, the lower-priority threads never execute.

To understand the effect that thread priorities have on context switch overhead, consider a three-thread environment with threads named *thread_1, thread_2*, and *thread_3*. Furthermore, assume that all the threads are suspended and waiting for a message. When *thread_1* receives a message, it immediately forwards it to *thread_2. Thread_2* then forwards the message to *thread_3. Thread_3* simply discards the message. After each thread processes its message, it suspends itself again and waits for another message. The processing required to execute these three threads varies greatly depending on their priorities. If all the threads have the same priority, a single context switch occurs between the execution of each thread. The context switch occurs when each thread suspends on an empty message queue.

However, if *thread_2* has higher priority than *thread_1* and *thread_3* has higher priority than *thread_2*, the number of context switches doubles. This is because another context switch occurs inside the tx_queue_send service when it detects that a higher-priority thread is now ready.

If distinct priorities for these threads are required, then the ThreadX preemption-threshold mechanism can prevent these extra context switches. This is an important feature because it allows several distinct thread priorities during thread scheduling, while at the same time eliminating some of the unwanted context switching that occurs during thread execution.

7.20.4 Consider Preemption-Threshold

Recall that a potential problem associated with thread priorities is priority inversion. Priority inversion occurs when a higher-priority thread becomes suspended because a lower-priority thread has a resource needed by the higher-priority thread. In some instances, it is necessary for two threads of different priority to share a common resource. If these threads are the only ones active, the priority inversion time is bounded by the time that the lower-priority thread holds the resource. This condition is both deterministic and quite normal. However, if one or more threads of an intermediate priority become active during this priority inversion condition—thus preempting the lower-priority thread—the priority inversion time becomes nondeterministic and the application may fail.

There are three primary methods of preventing priority inversion in ThreadX. First, the developer can select application priorities and design run-time behavior in a manner that prevents the priority inversion problem. Second, lower-priority threads can utilize preemption-threshold to block preemption from intermediate threads while they share resources with higher-priority threads. Finally, threads using ThreadX mutex objects to protect system resources may utilize the optional mutex priority inheritance to eliminate nondeterministic priority inversion.

7.20.5 Consider Priority Inheritance

In priority inheritance, a lower-priority thread temporarily acquires the priority of a higher-priority thread that is attempting to obtain the same mutex owned by the lower priority thread. When the lower-priority thread releases the mutex, its original priority is then restored and the higher-priority thread is given ownership of the mutex. This feature eliminates priority inversion by bounding the inversion time to the time the lower priority thread holds the mutex. Note that priority inheritance is available only with a mutex but not with a counting semaphore.

7.20.6 Consider Round-Robin Scheduling

ThreadX supports round-robin scheduling of multiple threads that have the same priority. This is accomplished through cooperative calls to the tx_thread_relinquish service. Calling this service gives all other ready threads with the same priority a chance to execute before the caller of the tx_thread_relinquish service executes again.

7.20.7 Consider Time-Slicing

Time-slicing provides another form of round-robin scheduling for threads with the same priority. ThreadX makes time-slicing available on a per-thread basis. An application assigns the thread's time-slice when it creates the thread and can modify the time-slice during run-time. When a thread's time-slice expires, all other ready threads with the same priority get a chance to execute before the time-sliced thread executes again.

7.21 Thread Internals

When the TX_THREAD data type is used to declare a thread, a TCB is created, and that TCB is added to a doubly linked circular list, as illustrated in Figure 7.18. The pointer named tx_thread_created_ptr points to the first TCB in the list. See the fields in the TCB for thread attributes, values, and other pointers.

When the scheduler needs to select a thread for execution, it chooses a ready thread with the highest priority. To determine such a thread, the scheduler first determines the priority

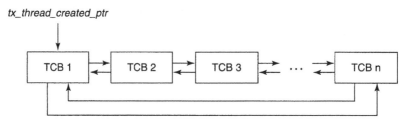

Figure 7.18: Created thread list

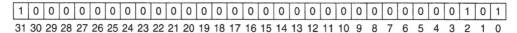

Figure 7.19: Map showing ready thread priorities

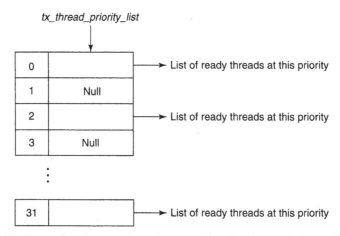

Figure 7.20: Example of array of ready thread head pointers indexed by priority

of the next ready thread. In order to get this priority, the scheduler consults the map of ready thread priorities. Figure 7.19 shows an example of this priority map.

This priority map (actually a ULONG variable) represents each of the 32 priorities with one bit. The bit pattern in the preceding example indicates that there are ready threads with priorities 0, 2, and 31.

When the highest priority of the next ready thread is determined, the scheduler uses an array of ready thread head pointers to select the thread to be executed. Figure 7.20 illustrates the organization of this array.

This array of TX_THREAD list-head pointers is directly indexed by thread priority. If an entry is non-NULL, there is at least one thread at that priority ready for execution. The threads in each priority list are managed in a doubly linked, circular list of TCBs, as illustrated in Figure 7.18. The thread in the front of the list represents the next thread to execute for that priority.

7.22 Overview

A thread is a dynamic entity that constitutes the basis for RTOS application development.

ThreadX supplies 13 services designed for a range of actions, including thread creation, deletion, modification, and termination.

The Thread Control Block (TCB) is a structure used to store vital information about a thread's state during execution. It also preserves vital thread information when a context switch becomes necessary.

Application developers have several options available when they create threads, including use of time-slicing and preemption-threshold. However, these options can be changed at any time during thread execution.

A thread can voluntarily relinquish control to another thread, it can resume execution of another thread, and it can abort the suspension of another thread.

A thread has five states: ready, suspended, executing, terminated, and completed. Only one thread executes at any point in time, but there may be many threads in the Ready Thread List and the Suspended Thread List. Threads in the former list are eligible for execution, while threads in the latter list are not.

7.23 Key Terms and Phrases

abort thread suspension	creating a thread
application timer	currently executing thread
auto start option	deleting a thread
change preemption-threshold	entry function input
change priority	executing state
change time-slice	execution context
completed state	interrupt service routine

cooperative multithreading	preemption-threshold
priority	Suspended thread list
ready state	suspension of thread
Ready thread List	terminate thread service
reentrant function	terminated state
relinquish processor control	Thread Control Block (TCB)
Resume thread execution	thread entry function
round-robin scheduling	thread execution state
Service return values	thread run-count
sleep mode	thread scheduling loop
stack size	thread start option
suspended state	time-slice

7.24 Problems

1. What is the value of tx_state (a member of the TCB) immediately after the service tx_thread_sleep(100); has been called?

2. What is the primary danger of specifying a thread stack size that is too small?

3. What is the primary danger of specifying a thread stack size that is too large?

4. Suppose tx_thread_create is called with a priority of 15, a preemption-threshold of 20, and a time-slice of 100. What would be the result of this service call?

5. Give an example showing how the use of reentrant functions is helpful in multithreaded applications.

6. Answer the following questions for this thread create sequence, where *&my_stack* is a valid pointer to the thread stack:

```
TX_THREAD my_thread;
UINT status;

    status = tx_thread_create(&my_thread, "my_thread",
                    my_thread_entry, 0x000F,
                    &my_stack, 2000, 25, 25,
                    150, TX_DONT_START);
```

    a   At what point in time will this thread be placed in the Ready Thread List?

    b   Is preemption-threshold used? If so, what is its value?

    c   Is time-slicing used? If so, what is its value?

    d   What is the size of the thread stack?

    e   What is the value of the variable *status* after this service is executed?

7   What is the difference between thread deletion and thread termination?

8   Given a pointer to any arbitrary thread, which thread service obtains the state of that thread and the number of times that thread has been executed?

9   If an executing thread has priority 15 and a preemption-threshold of 0, will another thread with priority 5 be able to preempt it?

10 Explain the difference between time-slicing and cooperative multithreading.

11 Under what circumstances will a thread be placed in the Ready Thread List? Under what circumstances will it be removed?

12 Under what circumstances will a thread be placed in the Suspended Thread List? Under what circumstances will it be removed?

13 Give an example in which using the tx_thread_wait_abort service would be essential.

14 How does the thread scheduling loop select a thread for execution?

15 Under what circumstances will a thread be preempted? What happens to that thread when it is preempted?

16 Describe the five states of a thread. Under what circumstances would a thread status be changed from executing to ready? From ready to suspended? From suspended to ready? From completed to ready?

CHAPTER 8

Mutual Exclusion Challenges and Considerations

8.1 Introduction

On many occasions, we need to guarantee that a thread has exclusive access to a shared resource or to a critical section. However, several threads may need to obtain these items, so we need to synchronize their behavior to ensure that exclusive access can be provided. In this chapter, we consider the properties of the mutex, which is designed solely to provide mutual exclusion protection by avoiding conflict between threads and preventing unwanted interactions between threads.

A mutex is a public resource that can be owned by, at most, one thread at any point in time. Furthermore, a thread (and only that same thread) can repeatedly[1] obtain the same mutex $2^{32}-1$ times, to be exact. However, that same thread (and only that thread) must give up that mutex the same number of times before the mutex becomes available again.

8.2 Protecting a Critical Section

A critical section is a code segment in which instructions must be executed in sequence without interruption. The mutex helps in achieving this goal. Consider Figure 8.1, which shows a code segment that is a critical section. To enter this critical section, a thread must first obtain ownership of a certain mutex that protects the critical section. Thus, when the thread is ready to begin executing this code segment, it first attempts to acquire that

---

[1]Some writers describe this type of mutex as a *recursive mutex* because of the same-thread, multiple-ownership capability. However, we will not use that terminology here.

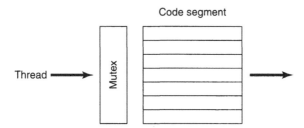

Figure 8.1: Mutex protecting a critical section

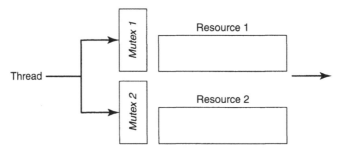

Figure 8.2: Mutexes providing exclusive access to multiple shared resources

mutex. After the thread has acquired the mutex, it executes the code segment, and then relinquishes the mutex.

8.3 Providing Exclusive Access to Shared Resources

A mutex can provide exclusive access to one shared resource in the same manner that it can protect a critical section. That is, a thread must first obtain the mutex before it can access the shared resource. However, if a thread must have exclusive access to two (or more) shared resources at the same time, then it must protect each shared resource with a separate mutex. In this case, the thread must first obtain a particular mutex for each of the shared resources before continuing. Figure 8.2 illustrates this process. When the thread is ready to access these resources, it first gets the two mutexes that protect these resources. After the thread has acquired both mutexes, it accesses the shared resources, and then relinquishes both mutexes after it has finished with these resources.

Field	Description
`tx_mutex_id`	Control block ID
`tx_mutex_name`	Pointer to mutex name
`tx_mutex_ownership_count`	Mutex ownership count
`*tx_mutex_owner`	Mutex ownership pointer
`tx_mutex_inherit`	Priority inheritance flag
`tx_mutex_original_priority`	Original priority of owning thread
`tx_mutex_original_threshold`	Original preemption-threshold of owning thread
`*tx_mutex_suspension_list`	Pointer to suspension list
`tx_mutex_suspended_count`	Suspension list count
`*tx_mutex_created_next`	Pointer to the next mutex in the created list
`*tx_mutex_created_previous`	Pointer to the previous mutex in the created list

Figure 8.3: Mutex Control Block

8.4 Mutex Control Block

The Mutex Control Block (MCB)[2] is a structure used to maintain the state of a mutex during run-time. It contains a variety of information, including the mutex owner, the ownership count, the priority inheritance flag, the original priority of the owning thread, the original preemption-threshold of the owning thread, the suspension count, and a pointer to the suspension list. Figure 8.3 contains many of the fields that comprise the MCB.

In most cases, the developer can ignore the contents of the MCB. However, in some situations, especially during debugging, inspecting certain members of the MCB is useful. Note that although ThreadX allows inspection of an MCB, it strictly prohibits modification of one.

8.5 Summary of Mutex Services

Appendix E contains detailed information about mutex services, providing the information on the following: prototype, brief description of the service, parameters,

---

[2]The characteristics of each mutex are contained in its MCB. This structure is defined in the tx_api.h file.

Mutex Service	Description
tx_mutex_create	Create a mutex
tx_mutex_delete	Delete a mutex
tx_mutex_get	Attempt to obtain ownership of a mutex
tx_mutex_info_get	Retrieve information about a mutex
tx_mutex_performance_info_get	Get mutex performance information
tx_mutex_performance_system_info_get	Get mutex system performance information
tx_mutex_prioritize	Put highest priority suspended thread at front of suspension list
tx_mutex_put	Release ownership of mutex

Figure 8.4: Mutex services

return values, notes and warnings, allowable invocation, preemption possibility, and an example that illustrates how the service can be used. Figure 8.4 contains a listing of all available mutex services. In the following sections of this chapter, you will study each of these services. We will consider the many features of the services, and we will develop an illustrative example of a sample system that uses them.

8.6 Creating a Mutex

A mutex is declared with the TX_MUTEX data type[3] and is defined with the tx_mutex_ create service. When defining a mutex, you need to specify the MCB, the name of the mutex, and the priority inheritance option. Figure 8.5 contains a list of these attributes. We will develop one example of mutex creation to illustrate the use of this service. We will give our mutex the name "my_mutex" and we will activate the priority inheritance feature. Priority inheritance allows a lower-priority thread to temporarily assume the priority of a higher-priority thread that is waiting for a mutex owned by the lower-priority thread. This feature helps the application to avoid priority inversion by eliminating preemption of intermediate thread priorities. Figure 8.6 contains an example of mutex creation.

If you wanted to create a mutex without the priority inheritance feature, you would use the TX_NO_INHERIT parameter rather than the TX_INHERIT parameter.

---

[3]When a mutex is declared, an MCB is created.

```
┌─────────────────────────────────┐
│      Mutex control block         │
├─────────────────────────────────┤
│         Mutex name               │
├─────────────────────────────────┤
│   Priority inheritance option    │
└─────────────────────────────────┘
```

Figure 8.5: Attributes of a mutex

```
TX_MUTEX my_mutex;
UINT status;

/* Create a mutex to provide protection over a
   shared resource. */
status = tx_mutex_create(&my_mutex,"my_mutex_name",
                         TX_INHERIT);

/* If status equals TX_SUCCESS, my_mutex is
   ready for use. */
```

Figure 8.6: Creating a mutex with priority inheritance

```
TX_MUTEX my_mutex;
UINT status;

…

/* Delete a mutex. Assume that the mutex
   has already been created. */
status = tx_mutex_delete(&my_mutex);

/* If status equals TX_SUCCESS, the mutex
   has been deleted. */
```

Figure 8.7: Deleting a mutex

8.7 Deleting a Mutex

A mutex can be deleted with the tx_mutex_delete service. When a mutex is deleted, all threads that have been suspended because they are waiting for that mutex are resumed (that is, placed on the Ready list). Each of these threads will receive a TX_DELETED return status from its call to tx_mutex_get. Figure 8.7 contains an example showing how the mutex called "my_mutex" can be deleted.

8.8 Obtaining Ownership of a Mutex

The tx_mutex_get service enables a thread to attempt to obtain exclusive ownership of a mutex. If no thread owns that mutex, then that thread acquires ownership of the mutex. If the calling thread already owns the mutex, then tx_mutex_get increments the ownership counter and returns a successful status. If another thread already owns the mutex, the action taken depends on the calling option used with tx_mutex_get, and whether the mutex has priority inheritance enabled. These actions are displayed in Figure 8.8.

tx_mutex_get Wait Option	Priority Inheritance Enabled in Mutex	Priority Inheritance Disabled in Mutex
TX_NO_WAIT	Immediate return	Immediate return
TX_WAIT_FOREVER	If the calling thread has a higher priority, the owning thread's priority is raised to that of the calling thread, then the calling thread is placed on the suspension list, and the calling thread waits indefinitely	Thread placed on suspension list and waits indefinitely
Timeout value	If the calling thread has a higher priority, the owning thread's priority is raised to that of the calling thread, then the calling thread is placed on the suspension list, and the calling thread waits until the number of specified timer-ticks has expired	Thread placed on suspension list and waits until the number of specified timer-ticks has expired

Figure 8.8: Actions taken when mutex is already owned by another thread

```
TX_MUTEX my_mutex;
UINT status;

...

/* Obtain exclusive ownership of the mutex "my_mutex".
   If the mutex called "my_mutex" is not available, suspend
   until it becomes available. */

status = tx_mutex_get(&my_mutex, TX_WAIT_FOREVER);
```

Figure 8.9: Obtain ownership of a mutex

If you use priority inheritance, make certain that you do not allow an external thread to modify the priority of the thread that has inherited a higher priority during mutex ownership. Figure 8.9 contains an example of a thread attempting to obtain ownership of a mutex.

If the variable *status* contains the value TX_SUCCESS, then this was a successful get operation. The TX_WAIT_FOREVER option was used in this example. Therefore, if the mutex is already owned by another thread, the calling thread will wait indefinitely in the suspension list.

8.9 Retrieving Mutex Information

There are three services that enable you to retrieve vital information about mutexes. The first such service for mutexes—the tx_mutex_info_get service—retrieves a subset of information from the Mutex Control Block. This information provides a "snapshot" at a particular instant in time, i.e., when the service is invoked. The other two services provide summary information that is based on the gathering of run-time performance data. One service—the tx_mutex_performance_info_get service—provides an information summary for a particular mutex up to the time the service is invoked. By contrast the tx_mutex_performance_system_ info_get retrieves an information summary for all mutexes in the system up to the time the service is invoked. These services are useful in analyzing the behavior of the system and determining whether there are potential problem areas. The tx_mutex_info_get[4] service obtains information that includes the ownership count, the location of the owning thread, the location of the first thread on the suspension list, the number of suspended threads, and the location of the next created mutex. Figure 8.10 shows how this service can be used.

If the variable *status* contains the value TX_SUCCESS, the information was successfully retrieved.

8.10 Prioritizing the Mutex Suspension List

When a thread is suspended because it is waiting for a mutex, it is placed in the suspension list in a FIFO manner. When the mutex becomes available, the first thread in the suspension list (regardless of priority) will obtain ownership of that mutex. The

---

[4]By default, only the *tx_mutex_info_get* service is enabled. The other two information-gathering services must be enabled in order to use them.

```
              TX_MUTEX my_mutex;
              CHAR *name;
              ULONG count;
              TX_THREAD *owner;
              TX_THREAD *first_suspended;
              ULONG suspended_count;
              TX_MUTEX *next_mutex;
              UINT status;

              ...

              /* Retrieve information about the previously
                 created mutex called "my_mutex." */

              status = tx_mutex_info_get(&my_mutex, &name,
                                         &count, &owner,
                                         &first_suspended,
                                         &suspended_count,
                                         &next_mutex);
              /* If status equals TX_SUCCESS, the information
                 requested is valid. */
```

Figure 8.10: Example showing how to retrieve mutex information

```
    TX_MUTEX my_mutex;
    UINT status;

    ...

    /* Ensure that the highest priority thread will receive
       ownership of the mutex when it becomes available. */
    status = tx_mutex_prioritize(&my_mutex);

    /* If status equals TX_SUCCESS, the highest priority
       suspended thread has been placed at the front of the
       list. The next tx_mutex_put call that releases
       ownership of the mutex will give ownership to this
       thread and wake it up. */
```

Figure 8.11: Prioritizing the mutex suspension list

tx_mutex_prioritize service places the highest-priority thread suspended for ownership of a specific mutex at the front of the suspension list. All other threads remain in the same FIFO order in which they were suspended. Figure 8.11 shows how this service can be used.

```
TX_MUTEX my_mutex;
UINT status;

...

/* Release ownership of "my_mutex." */
status = tx_mutex_put(&my_mutex);

/* If status equals TX_SUCCESS, the mutex ownership
   count has been decremented and if zero, released. */
```

Figure 8.12: Releasing ownership of a mutex

If the variable *status* contains the value TX_SUCCESS, the highest-priority thread in the suspension list that is waiting for the mutex called "my_mutex" has been placed at the front of the suspension list. If no thread was waiting for this mutex, the return value is also TX_SUCCESS and the suspension list remains unchanged.

8.11 Releasing Ownership of a Mutex

The tx_mutex_put service enables a thread to release ownership of a mutex. Assuming that the thread owns the mutex, the ownership count is decremented. If the ownership count becomes zero, the mutex becomes available. If the mutex becomes available and if priority inheritance is enabled for this mutex, then the priority of the releasing thread reverts to the priority it had when it originally obtained ownership of the mutex. Any other priority changes made to the releasing thread during ownership of the mutex may be undone also. Figure 8.12 shows how this service can be used.

If the variable *status* contains the value TX_SUCCESS, then the put operation was successful, and the ownership count was decremented.

8.12 Avoiding the Deadly Embrace

One of the potential pitfalls in using mutexes[5] is the so-called *deadly embrace*. This is an undesirable situation in which two or more threads become suspended indefinitely while attempting to get mutexes already owned by other threads. Figure 8.13 illustrates a scenario that leads to a deadly embrace. Following is the sequence of events depicted in this figure.

---

[5]This problem is also associated with the use of semaphores, which we discuss in Chapter 11.

1. *Thread 1* obtains ownership of *Mutex 1*

2. *Thread 2* obtains ownership of *Mutex 2*

3. *Thread 1* suspends because it attempts to obtain ownership of *Mutex 2*

4. *Thread 2* suspends because it attempts to obtain ownership of *Mutex 1*

Thus, *Thread 1* and *Thread 2* have entered a deadly embrace because they have suspended indefinitely, each waiting for the mutex that the other thread owns.

How can you avoid deadly embraces? Prevention at the application level is the only method for real-time systems. The only way to guarantee the absence of deadly embraces is to permit a thread to own at most one mutex at any time. If threads must own multiple mutexes, you can generally avoid deadly embraces if you make the threads gather the mutexes in the same order. For example, the deadly embrace in Figure 8.13 could be prevented if the threads would always obtain the two mutexes in consecutive order, i.e., *Thread 1* (or *Thread 2*) would attempt to acquire *Mutex 1*, and then would immediately attempt to acquire *Mutex 2*. The other thread would attempt to acquire *Mutex 1* and *Mutex 2* in the same order.

One way to recover from a deadly embrace is to use the suspension time-out feature associated with the tx_mutex_get service, which is one of the three available wait

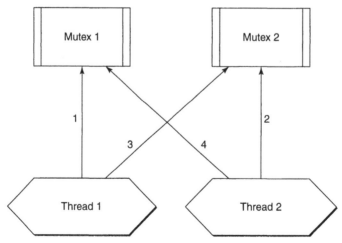

Figure 8.13: Sequence of actions leading to a deadly embrace

options. Another way to recover from a deadly embrace is for another thread to invoke the tx_thread_wait_abort service to abort the suspension of a thread trapped in a deadly embrace.

8.13 Sample System Using a Mutex to Protect Critical Sections

We will create a sample system to illustrate how a mutex can be used to protect the critical sections of two threads. This system was introduced in Chapter 2 where Speedy_Thread and Slow_Thread each had four activities, two of which were critical sections. Figure 8.14 and Figure 8.15 show the sequence of activities for each of these two threads, where the shaded boxes represent the critical sections.

In order to develop this system, we will need to create two threads and one mutex. Each thread must have its own stack, which we will implement as an array, rather than as a memory byte pool. We will need to create the thread entry functions that will perform the desired activities. Because we will create this system in its entirety, we outline this process with Figure 8.16, which is a variation of the basic four-part system structure that first appeared in Chapter 2.

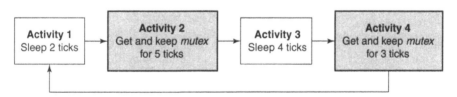

Figure 8.14: Activities of the Speedy_Thread (priority = 5)

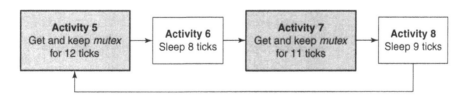

Figure 8.15: Activities of the Slow_Thread (priority = 15)

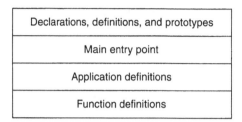

Figure 8.16: Basic system structure

For the first part of the system, we declare as global entities the two threads, the one mutex, and the two thread stacks as follows:

```
TX_THREAD Speedy_Thread, Slow_Thread;
TX_MUTEX my_mutex;
#DEFINE STACK_SIZE 1024;
CHAR stack_speedy [STACK_SIZE], stack_slow[STACK_SIZE];
```

The process of declaring the threads creates two Thread Control Blocks (TCBs), and declaring the mutex creates its MCB as well. The thread stacks will be ready for use in the tx_application_define function.

The second part of the system is where we define the main entry point, which is the call to enter the ThreadX kernel.

The third part of the system is where we define the threads and the mutex. Following is the definition of Speedy_Thread.

```
tx_thread_create (&Speedy_Thread, "Speedy_Thread",
            Speedy_Thread_entry, 0,
            stack_speedy, STACK_SIZE,
            5, 5, TX_NO_TIME_SLICE, TX_AUTO_START);
```

Speedy_Thread has a priority of 5, but does not have a preemption-threshold, nor does it have a time-slice. Following is the definition of Slow_Thread.

```
tx_thread_create (&Slow_Thread, "Slow_Thread",
          Slow_Thread_entry, 1,
          stack_slow, STACK_SIZE,
          15, 15, TX_NO_TIME_SLICE, TX_AUTO_START);
```

Slow_Thread has a priority of 15, but does not have a preemption-threshold, nor does it have a time-slice. Both threads will start immediately. Following is the definition of *my_mutex*.

```
tx_mutex_create(&my_mutex, "my_mutex", TX_NO_INHERIT);
```

The mutex is given a name but does not have the priority inheritance feature.

The fourth part of our system is where we develop the thread entry functions. Following is a portion of the entry function for the Speedy_Thread.

```
/* Activity 1: 2 timer-ticks. */
     tx_thread_sleep(2);
/* Activity 2—critical section—5 timer-ticks
     Get the mutex with suspension */
tx_mutex_get(&my_mutex, TX_WAIT_FOREVER);
     tx_thread_sleep(5);
/* Release the mutex */
tx_mutex_put(&my_mutex);
```

The first two activities of Speedy_Thread are represented here. Activity 1 is not a critical section, so we immediately sleep for two timer-ticks. Activity 2 is a critical section, so to execute it we must first obtain ownership of the mutex. After we get the mutex, we sleep for five timer-ticks. The other activities for both threads follow a similar pattern. When we develop the complete system, we will check the status of the return values to make certain the service calls have been performed correctly.

Figure 8.17 through Figure 8.21 contain a complete listing for this sample system, separated into five parts, where the last two parts are the thread entry functions. The complete program listing called *08_sample_system.c* is located in a later section of this chapter and on the enclosed CD.

The first part of the sample system contains all the necessary directives, declarations, definitions, and prototypes.

```
/* 08_sample_system.c
   Create two threads, and one mutex.
   Use an array for the thread stacks.
   The mutex protects the critical sections.  */

#include   "tx_api.h"
#include   <stdio.h>

#define     STACK_SIZE           1024

CHAR stack_speedy[STACK_SIZE];
CHAR stack_slow[STACK_SIZE];

/* Define the ThreadX object control blocks... */

TX_THREAD                Speedy_Thread;
TX_THREAD                Slow_Thread;

TX_MUTEX                 my_mutex;

/* Define thread prototypes.  */

void    Speedy_Thread_entry(ULONG thread_input);
void    Slow_Thread_entry(ULONG thread_input);
```

Figure 8.17: Definitions, declarations, and prototypes

```
/* Define main entry point.  */

int main()
{
    /* Enter the ThreadX kernel.  */
    tx_kernel_enter();
}
```

Figure 8.18: The main entry point

The second part of the sample system contains the main entry point. This is the entry into the ThreadX kernel. Note that the call to tx_kernel_enter does not return, so do not place any processing after it.

The third part of the sample system consists of the application definition function called tx_application_define. This function can be used to define all the application resources in the system. This function has a single input parameter, which is the first available RAM

```
/* Define what the initial system looks like.  */

void    tx_application_define(void *first_unused_memory)
{

    /* Put system definitions here,
       e.g., thread and mutex creates */

    /* Create the Speedy_Thread.  */
    tx_thread_create(&Speedy_Thread, "Speedy_Thread",
                     Speedy_Thread_entry, 0,
                     stack_speedy, STACK_SIZE,
                     5, 5, TX_NO_TIME_SLICE, TX_AUTO_START);
    /* Create the Slow_Thread */
    tx_thread_create(&Slow_Thread, "Slow_Thread",
                     Slow_Thread_entry, 1,
                     stack_slow, STACK_SIZE,
                     15, 15, TX_NO_TIME_SLICE, TX_AUTO_START);
    /* Create the mutex used by both threads  */
    tx_mutex_create(&my_mutex, "my_mutex", TX_NO_INHERIT);
}
```

Figure 8.19: Application definitions

address. This is typically used as a starting point for initial run-time memory allocations of thread stacks, queues, and memory pools.

The fourth part of the sample system consists of the entry function for the Speedy_ Thread. This function defines the four activities of the thread, and displays the current time each time the thread finishes a complete cycle.

The fifth and final part of the sample system consists of the entry function for the Slow_ Thread. This function defines the four activities of the thread, and displays the current time each time the thread finishes a complete cycle.

8.14 Output Produced by Sample System

Figure 8.22 contains some output produced by executing the sample system for a few thread activity cycles. Your output should be similar, but not necessarily identical.

The minimum amount of time that the Speedy_Thread requires to complete its cycle of activities is 14 timer-ticks. By contrast, the Slow_Thread requires at least 40 timer-ticks

```
/* Define the activities for the Speedy_Thread */

void    Speedy_Thread_entry(ULONG thread_input)
{
UINT    status;
ULONG   current_time;

   while(1)
   {
      /* Activity 1:  2 timer-ticks. */
      tx_thread_sleep(2);

      /* Activity 2 - critical section - 5 timer-ticks
      Get the mutex with suspension. */

      status = tx_mutex_get(&my_mutex, TX_WAIT_FOREVER);
      if (status != TX_SUCCESS)  break;  /* Check status */

      tx_thread_sleep(5);

      /* Release the mutex. */
      status = tx_mutex_put(&my_mutex);
      if (status != TX_SUCCESS)  break;  /* Check status */

      /* Activity 3:  4 timer-ticks. */
      tx_thread_sleep(4);

      /* Activity 4- critical section - 3 timer-ticks
      Get the mutex with suspension. */

      status = tx_mutex_get(&my_mutex, TX_WAIT_FOREVER);
      if (status != TX_SUCCESS)  break;  /* Check status */

      tx_thread_sleep(3);

      /* Release the mutex. */
      status = tx_mutex_put(&my_mutex);
      if (status != TX_SUCCESS)  break;  /* Check status */

      current_time = tx_time_get();
      printf("Current Time:  %lu  Speedy_Thread finished cycle...\n",
             current_time);

   }

}
```

Figure 8.20: Speedy_Thread entry function

```
/* Define the activities for the Slow_Thread */

void    Slow_Thread_entry(ULONG thread_input)
{
UINT    status;
ULONG   current_time;

    while(1)
    {
        /* Activity 5 - critical section - 12 timer-ticks
        Get the mutex with suspension. */

        status = tx_mutex_get(&my_mutex, TX_WAIT_FOREVER);
        if (status != TX_SUCCESS)  break;  /* Check status */

        tx_thread_sleep(12);

        /* Release the mutex. */
        status = tx_mutex_put(&my_mutex);
        if (status != TX_SUCCESS)  break;  /* Check status */

        /* Activity 6:  8 timer-ticks. */
        tx_thread_sleep(8);

        /* Activity 7 - critical section - 11 timer-ticks
        Get the mutex with suspension. */

        status = tx_mutex_get(&my_mutex, TX_WAIT_FOREVER);
        if (status != TX_SUCCESS)  break;  /* Check status */

        tx_thread_sleep(11);

        /* Release the mutex. */
        status = tx_mutex_put(&my_mutex);
        if (status != TX_SUCCESS)  break;  /* Check status */

        /* Activity 8:  9 timer-ticks. */
        tx_thread_sleep(9);

        current_time = tx_time_get();
        printf("Current Time:  %lu  Slow_Thread finished cycle...\n",
                current_time);
    }
}
```

Figure 8.21: Slow_Thread entry function

```
Current Time:      34   Speedy_Thread finished cycle...
Current Time:      40   Slow_Thread finished cycle...
Current Time:      56   Speedy_Thread finished cycle...
Current Time:      77   Speedy_Thread finished cycle...
Current Time:      83   Slow_Thread finished cycle...
Current Time:      99   Speedy_Thread finished cycle...
Current Time:     120   Speedy_Thread finished cycle...
Current Time:     126   Slow_Thread finished cycle...
Current Time:     142   Speedy_Thread finished cycle...
Current Time:     163   Speedy_Thread finished cycle...
```

Figure 8.22: Some output produced by sample system

to complete one cycle of its activities. However, the critical sections of the Slow_Thread will cause delays for the Speedy_Thread. Consider the sample output in Figure 8.22, in which the Speedy_Thread finishes its first cycle at time 34, meaning that it encountered a delay of 20 timer-ticks because of the Slow_Thread. The Speedy_Thread completes subsequent cycles in a more timely fashion but it will always spend a lot of time waiting for the Slow_Thread to complete its critical section.

To better understand what is happening with the sample system, let us trace a few actions that occur. After initialization has been completed, both threads are on the Ready Thread List and are ready to execute. The scheduler selects Speedy_Thread for execution because it has a higher priority than Slow_Thread. Speedy_Thread begins Activity 1, which causes it to sleep two timer-ticks, i.e., it is placed on the Suspend Thread List during this time. Slow_Thread then gets to execute and it begins Activity 5, which is a critical section. Slow_Thread takes ownership of the mutex and goes to sleep for 12 times timer-ticks, i.e., it is placed in the Suspend Thread List during this time. At time 2, Speedy_Thread is removed from the Suspend Thread List, placed on the Ready Thread List, and begins Activity 2, which is a critical section. Speedy_Thread attempts to obtain ownership of the mutex, but it is already owned, so Speedy_Thread is placed in the Suspend Thread List until the mutex is available. At time 12, Slow_Thread is placed back in the Ready Thread List and gives up ownership of the mutex. Figure 8.23 contains a partial trace of the actions for the sample system.

8.15 Listing for 08_sample_system.c

The sample system named *08_sample_system.c* is located on the attached CD. The complete listing appears below; line numbers have been added for easy reference.

Time	Actions performed	Mutex owner
Initial	*Speedy* and *Slow* on Thread Ready List (TRL), Thread Suspension List (TSL) empty	none
0	*Speedy* sleeps 2, placed on TSL, *Slow* takes mutex, sleeps 12, placed on TSL	*Slow*
2	*Speedy* wakes up, put on TRL, unable to get mutex, placed on TSL	*Slow*
12	*Slow* wakes up, put on TRL, gives up mutex, *Speedy* preempts *Slow*, *Speedy* takes mutex, sleeps 5, put on TSL, *Slow* sleeps 8, put on TSL	*Speedy*
17	*Speedy* wakes up, put on TRL, gives up mutex, sleeps 4, put on TSL	none
20	*Slow* wakes up, put on TRL, takes mutex, sleeps 11, put on TSL	*Slow*
21	*Speedy* wakes up, put on TRL, unable to get mutex, put on TSL	*Slow*
31	*Slow* wakes up, put on TRL, gives up mutex, *Speedy* preempts *Slow*, *Speedy* takes mutex, sleeps 3, put on TSL, *Slow* sleeps 9, put on TSL	*Speedy*
34	*Speedy* wakes up, put on TRL, gives up mutex, sleeps 3, put on TSL (this completes one full cycle for *Speedy*)	none
37	*Speedy* wakes up, put on TRL, sleeps 2, put on TSL	none
39	*Speedy* wakes up, put on TRL, takes mutex, sleeps 5, put on TSL	*Speedy*
40	*Slow* wakes up, put on TRL, unable to get mutex, put on TSL (this completes one full cycle for *Slow*)	*Speedy*

Figure 8.23: Partial activity trace of sample system

```
001  /* 08_sample_system.c
002
003  Create two threads, and one mutex.
004  Use an array for the thread stacks.
005  The mutex protects the critical sections. */
006
```

```
007   /*****************************************************/
008   /*       Declarations, Definitions, and Prototypes      */
009   /*****************************************************/
010
011   #include "tx_api.h"
012   #include <stdio.h >
013
014   #define    STACK_SIZE       1024
015
016   CHAR stack_speedy[STACK_SIZE];
017   CHAR stack_slow[STACK_SIZE];
018
019
020   /* Define the ThreadX object control blocks... */
021
022   TX_THREAD     Speedy_Thread;
023   TX_THREAD     Slow_Thread;
024
025   TX_MUTEX      my_mutex;
026
027
028   /* Define thread prototypes. */
029
030   void    Speedy_Thread_entry(ULONG thread_input);
031   void    Slow_Thread_entry(ULONG thread_input);
032
033
034   /*****************************************************/
035   /*             Main Entry Point                      */
036   /*****************************************************/
037
038   /* Define main entry point. */
039
040   int main()
041   {
042
043       /* Enter the ThreadX kernel. */
044       tx_kernel_enter();
045   }
046
```

```
047
048   /***************************************************/
049   /*              Application Definitions            */
050   /***************************************************/
051
052   /* Define what the initial system looks like. */
053
054   void tx_application_define(void *first_unused_memory)
055   {
056
057
058       /* Put system definitions here,
059           e.g., thread and mutex creates */
060
061       /* Create the Speedy_Thread. */
062       tx_thread_create(&Speedy_Thread, "Speedy_Thread",
063                       Speedy_Thread_entry, 0,
064                       stack_speedy, STACK_SIZE,
065                       5, 5, TX_NO_TIME_SLICE, TX_AUTO_START);
066
067       /* Create the Slow_Thread */
068       tx_thread_create(&Slow_Thread, "Slow_Thread",
069                       Slow_Thread_entry, 1,
070                       stack_slow, STACK_SIZE,
071                       15, 15, TX_NO_TIME_SLICE, TX_AUTO_START);
072
073       /* Create the mutex used by both threads */
074       tx_mutex_create(&my_mutex, "my_mutex", TX_NO_INHERIT);
075
076   }
077
078
079   /***************************************************/
080   /*              Function Definitions               */
081   /***************************************************/
082
083   /* Define the activities for the Speedy_Thread */
084
085   void Speedy_Thread_entry(ULONG thread_input)
```

```
086  {
087  UINT status;
088  ULONG current_time;
089
090      while(1)
091      {
092
093          /* Activity 1: 2 timer-ticks. */
094          tx_thread_sleep(2);
095
096          /* Activity 2: 5 timer-ticks *** critical section ***
097          Get the mutex with suspension. */
098
099          status = tx_mutex_get(&my_mutex, TX_WAIT_FOREVER);
100          if (status ! = TX_SUCCESS) break; /* Check status */
101
102          tx_thread_sleep(5);
103
104          /* Release the mutex. */
105          status = tx_mutex_put(&my_mutex);
106          if (status ! = TX_SUCCESS) break; /* Check status */
107
108          /* Activity 3: 4 timer-ticks. */
109          tx_thread_sleep(4);
110
111          /* Activity 4: 3 timer-ticks *** critical section ***
112          Get the mutex with suspension. */
113
114          status = tx_mutex_get(&my_mutex, TX_WAIT_FOREVER);
115          if (status ! = TX_SUCCESS) break; /* Check status */
116
117          tx_thread_sleep(3);
118
119          /* Release the mutex. */
120          status = tx_mutex_put(&my_mutex);
121          if (status ! = TX_SUCCESS) break; /* Check status */
122
123          current_time = tx_time_get();
124          printf("Current Time: %lu Speedy_Thread finished
                     cycle...\n",
```

```
125                      current_time);
126
127      }
128  }
129
130  /****************************************************/
131
132  /* Define the activities for the Slow_Thread */
133
134  void Slow_Thread_entry(ULONG thread_input)
135  {
136  UINT status;
137  ULONG current_time;
138
139     while(1)
140     {
141
142        /* Activity 5: 12 timer-ticks *** critical section ***
143        Get the mutex with suspension. */
144
145        status = tx_mutex_get(&my_mutex, TX_WAIT_FOREVER);
146        if (status != TX_SUCCESS) break; /* Check status */
147
148        tx_thread_sleep(12);
149
150        /* Release the mutex. */
151        status = tx_mutex_put(&my_mutex);
152        if (status != TX_SUCCESS) break; /* Check status */
153
154        /* Activity 6: 8 timer-ticks. */
155        tx_thread_sleep(8);
156
157        /* Activity 7: 11 timer-ticks *** critical section ***
158        Get the mutex with suspension. */
159
160        status = tx_mutex_get(&my_mutex, TX_WAIT_FOREVER);
161        if (status != TX_SUCCESS) break; /* Check status */
162
163        tx_thread_sleep(11);
164
```

```
165          /* Release the mutex. */
166          status = tx_mutex_put(&my_mutex);
167          if (status ! = TX_SUCCESS) break; /* Check status */
168
169          /* Activity 8: 9 timer-ticks. */
170          tx_thread_sleep(9);
171
172          current_time = tx_time_get();
173          printf("Current Time: %lu Slow_Thread finished cycle...\n",
174                    current_time);
175
176      }
177  }
```

8.16 Mutex Internals

When the TX_MUTEX data type is used to declare a mutex, an MCB is created, and that MCB is added to a doubly linked circular list, as illustrated in Figure 8.24.

The pointer named tx_mutex_created_ptr points to the first MCB in the list. See the fields in the MCB for mutex attributes, values, and other pointers.

If the priority inheritance feature has been specified (i.e., the MCB field named tx_mutex_inherit has been set), the priority of the owning thread will be increased to match that of a thread with a higher priority that suspends on this mutex. When the owning thread releases the mutex, then its priority is restored to its original value, regardless of any intermediate priority changes. Consider Figure 8.25, which contains a sequence of operations for the thread named *my_thread* with priority 25, which successfully obtains the mutex named *my_mutex*, which has the priority inheritance feature enabled.

The thread called *my_thread* had an initial priority of 25, but it inherited a priority of 10 from the thread called *big_thread*. At this point, *my_thread* changed its own priority twice (perhaps unwisely because it lowered its own priority!). When *my_thread* released the mutex, its priority reverted to its original value of 25, despite the intermediate priority changes. Note that if *my_thread* had previously specified a preemption threshold, then the new preemption-threshold value would be changed to the new priority when a change-priority operation was executed. When *my_thread* released the mutex, then the

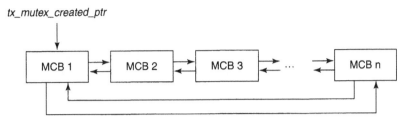

tx_mutex_created_ptr

Figure 8.24: Created mutex list

Action	Priority of *my_thread*
my_thread obtains *my_mutex*	25
big_thread (priority = 10) attempts to obtain *my_mutex*, but is suspended because the mutex is owned by *my_thread*	10
my_thread changes its own priority to 15	15
my_thread changes its own priority to 21	21
my_thread releases *my_mutex*	25

Figure 8.25: Example showing effect of priority inheritance on thread priority

preemption threshold would be changed to the original priority value, rather than to the original preemption-threshold value.

8.17 Overview

A mutex is a public resource that can be owned by at most one thread at any point in time. It has only one purpose: to provide exclusive access to a critical section or to shared resources.

Declaring a mutex has the effect of creating an MCB, which is a structure used to store vital information about that mutex during execution.

There are eight services designed for a range of actions involving mutexes, including creating a mutex, deleting a mutex, prioritizing a suspension list, obtaining ownership of a mutex, retrieving mutex information (3), and relinquishing ownership of a mutex.

Developers can specify a priority inheritance option when defining a mutex, or during later execution. Using this option will diminish the problem of priority inversion.

Another problem associated with the use of mutexes is the deadly embrace, and several tips for avoiding this problem were presented.

We developed a complete system that employs two threads and one mutex that protects the critical section of each thread. We presented and discussed a partial trace of the threads.

8.18 Key Terms and Phrases

creating a mutex	ownership of mutex
critical section	prioritize mutex suspension list
deadly embrace	priority inheritance
deleting a mutex	priority inversion
exclusive access	recovery from deadly embrace
multiple mutex ownership	shared resources
mutex	synchronize thread behavior
Mutex Control Block (MCB)	Ready Thread List
mutex wait options	Suspend Thread List
mutual exclusion	

8.19 Problems

1. Describe precisely what happens as a result of the following mutex declaration: TX_MUTEX mutex_1;

2. What is the difference between a mutex declaration and a mutex definition?

3. Suppose that a mutex is not owned, and a thread acquires that mutex with the tx_mutex_get service. What is the value of tx_mutex_suspended_count (a member of the MCB) immediately after that service has completed?

4. Suppose a thread with the lowest possible priority owns a certain mutex, and a ready thread with the highest possible priority needs that mutex. Will the high priority thread be successful in taking that mutex from the low-priority thread?

5. Describe all the circumstances (discussed so far) that would cause an executing thread to be moved to the Suspend Thread List.

6. Suppose a mutex has the priority-inheritance option enabled and a thread that attempted to acquire that mutex had its priority raised as a result. Exactly when will that thread have its priority restored to its original value?

7. Is it possible for the thread in the previous problem to have its priority changed while it is in the Suspend Thread List? If so, what are the possible problems that might arise? Are there any circumstances that might justify performing this action?

8. Suppose you were charged with the task of creating a watchdog thread that would try to detect and correct deadly embraces. Describe, in general terms, how you would accomplish this task.

9. Describe the purpose of the tx_mutex_prioritize service, and give an example.

10. Discuss two ways in which you can help avoid the priority inversion problem.

11. Discuss two ways in which you can help avoid the deadly embrace problem.

12. Consider Figure 8.23, which contains a partial activity trace of the sample system. Exactly when will the Speedy_Thread preempt the Slow_Thread?

CHAPTER 9

Memory Management: Byte Pools and Block Pools

9.1 Introduction

Recall that we used arrays for the thread stacks in the previous chapter. Although this approach has the advantage of simplicity, it is frequently undesirable and is quite inflexible. This chapter focuses on two ThreadX memory management resources that provide a good deal of flexibility: memory byte pools and memory block pools.

A *memory byte pool* is a contiguous block of bytes. Within such a pool, byte groups of any size (subject to the total size of the pool) may be used and reused. Memory byte pools are flexible and can be used for thread stacks and other resources that require memory. However, this flexibility leads to some problems, such as fragmentation of the memory byte pool as groups of bytes of varying sizes are used.

A *memory block pool* is also a contiguous block of bytes, but it is organized into a collection of fixed-size memory blocks. Thus, the amount of memory used or reused from a memory block pool is always the same—the size of one fixed-size memory block. There is no fragmentation problem, and allocating and releasing memory blocks is fast. In general, the use of memory block pools is preferred over memory byte pools.

We will study and compare both types of memory management resources in this chapter. We will consider the features, capabilities, pitfalls, and services for each type. We will also create illustrative sample systems using these resources.

9.2 Summary of Memory Byte Pools

A memory byte pool is similar to a standard C heap.[1] In contrast to the C heap, a ThreadX application may use multiple memory byte pools. In addition, threads can suspend on a memory byte pool until the requested memory becomes available.

Allocations from memory byte pools resemble traditional *malloc* calls, which include the amount of memory desired (in bytes). ThreadX allocates memory from the memory byte pool in a *first-fit* manner, i.e., it uses the first free memory block that is large enough to satisfy the request. ThreadX converts excess memory from this block into a new block and places it back in the free memory list. This process is called *fragmentation*.

When ThreadX performs a subsequent allocation search for a large-enough block of free memory, it merges adjacent free memory blocks together. This process is called *defragmentation*.

Each memory byte pool is a public resource; ThreadX imposes no constraints on how memory byte pools may be used.[2] Applications may create memory byte pools either during initialization or during run-time. There are no explicit limits on the number of memory byte pools an application may use.

The number of allocatable bytes in a memory byte pool is slightly less than what was specified during creation. This is because management of the free memory area introduces some overhead. Each free memory block in the pool requires the equivalent of two C pointers of overhead. In addition, when the pool is created, ThreadX automatically divides it into two blocks, a large free block and a small permanently allocated block at the end of the memory area. This allocated end block is used to improve performance of the allocation algorithm. It eliminates the need to continuously check for the end of the pool area during merging. During run-time, the amount of overhead in the pool typically increases. This is partly because when an odd number of bytes is allocated, ThreadX pads out the block to ensure proper alignment of the next memory block. In addition, overhead increases as the pool becomes more fragmented.

---

[1]In C, a heap is an area of memory that a program can use to store data in variable amounts that will not be known until the program is running.

[2]However, memory byte pool services cannot be called from interrupt service routines. (This topic will be discussed in a later chapter.)

The memory area for a memory byte pool is specified during creation. Like other memory areas, it can be located anywhere in the target's address space. This is an important feature because of the considerable flexibility it gives the application. For example, if the target hardware has a high-speed memory area and a low-speed memory area, the user can manage memory allocation for both areas by creating a pool in each of them.

Application threads can suspend while waiting for memory bytes from a pool. When sufficient contiguous memory becomes available, the suspended threads receive their requested memory and are resumed. If multiple threads have suspended on the same memory byte pool, ThreadX gives them memory and resumes them in the order they occur on the Suspended Thread List (usually FIFO). However, an application can cause priority resumption of suspended threads, by calling tx_byte_pool_prioritize prior to the byte release call that lifts thread suspension. The byte pool prioritize service places the highest priority thread at the front of the suspension list, while leaving all other suspended threads in the same FIFO order.

9.3 Memory Byte Pool Control Block

The characteristics of each memory byte pool are found in its Control Block.[3] It contains useful information such as the number of available bytes in the pool. Memory Byte Pool Control Blocks can be located anywhere in memory, but it is most common to make the Control Block a global structure by defining it outside the scope of any function. Figure 9.1 contains many of the fields that comprise this Control Block.

In most cases, the developer can ignore the contents of the Memory Byte Pool Control Block. However, there are several fields that may be useful during debugging, such as the number of available bytes, the number of fragments, and the number of threads suspended on this memory byte pool.

9.4 Pitfalls of Memory Byte Pools

Although memory byte pools provide the most flexible memory allocation, they also suffer from somewhat nondeterministic behavior. For example, a memory byte pool may have 2,000 bytes of memory available but not be able to satisfy an allocation request of

---

[3]The structure of the Memory Byte Pool Control Block is defined in the **tx_api.h** file.

Field	Description
tx_byte_pool_id	Byte pool ID
tx_byte_pool_name	Pointer to byte pool name
tx_byte_pool_available	Number of available bytes
tx_byte_pool_fragments	Number of fragments in the pool
tx_byte_pool_list	Head pointer of the byte pool
tx_byte_pool_search	Pointer for searching for memory
tx_byte_pool_start	Starting address of byte pool area
tx_byte_pool_size	Byte pool size (in bytes)
*tx_byte_pool_owner	Pointer to owner of a byte pool during a search
*tx_byte_pool_suspension_list	Byte pool suspension list head
tx_byte_pool_suspended_count	Number of threads suspended
*tx_byte_pool_created_next	Pointer to the next byte pool in the created list
*tx_byte_pool_created_previous	Pointer to the previous byte pool in the created list

Figure 9.1: Memory Byte Pool Control Block

even 1,000 bytes. This is because there is no guarantee on how many of the free bytes are contiguous. Even if a 1,000-byte free block exists, there is no guarantee on how long it might take to find the block. The allocation service may well have to search the entire memory pool to find the 1,000-byte block. Because of this problem, it is generally good practice to avoid using memory byte services in areas where deterministic, real-time behavior is required. Many such applications pre-allocate their required memory during initialization or run-time configuration. Another option is to use a memory block pool (discussed later in this chapter).

Users of byte pool allocated memory must not write outside its boundaries. If this happens, corruption occurs in an adjacent (usually subsequent) memory area. The results are unpredictable and quite often catastrophic.

9.5 Summary of Memory Byte Pool Services

Appendix B contains detailed information about memory byte pool services. This appendix contains information about each service, such as the prototype, a brief description of the service, required parameters, return values, notes and warnings, allowable invocation, and an example showing how the service can be used. Figure 9.2 contains a listing of all available memory byte pool services. In the subsequent sections of this chapter, we will investigate each of these services.

Memory byte pool service	Description
tx_byte_allocate	Allocate bytes of memory
tx_byte_pool_create	Create a memory byte pool
tx_byte_pool_delete	Delete a memory byte pool
tx_byte_pool_info_get	Retrieve information about the memory byte pool
tx_byte_pool_performance_info_get	Get byte pool performance information
tx_byte_pool_performance_system_info_get	Get byte pool system performance information
tx_byte_pool_prioritize	Prioritize the memory byte pool suspension list
tx_byte_release	Release bytes back to the memory byte pool

Figure 9.2: Services of the memory byte pool

We will first consider the tx_byte_pool_create service because it must be invoked before any of the other services.

9.6 Creating a Memory Byte Pool

A memory byte pool is declared with the TX_BYTE_POOL data type and is defined with the tx_byte_pool_create service. When defining a memory byte pool, you need to specify its Control Block, the name of the memory byte pool, the address of the memory byte pool, and the number of bytes available. Figure 9.3 contains a list of these attributes.

We will develop one example of memory byte pool creation to illustrate the use of this service. We will give our memory byte pool the name "my_pool." Figure 9.4 contains an example of memory byte pool creation.

If variable *status* contains the return value TX_SUCCESS, then a memory byte pool called my_pool that contains 2,000 bytes, and which begins at location 0×500000 has been created successfully.

9.7 Allocating from a Memory Byte Pool

After a memory byte pool has been declared and defined, we can start using it in a variety of applications. The tx_byte_allocate service is the method by which bytes of memory

Memory byte pool control block
Memory byte pool name
Location of memory byte pool
Total number of bytes available for memory byte pool

Figure 9.3: Attributes of a memory byte pool

```
UINT status;
TX_BYTE_POOL my_pool;

/* Create a memory pool whose total size is 2000 bytes
   starting at address 0x500000. */

status = tx_byte_pool_create(&my_pool, "my_pool",
                             (VOID *) 0x500000, 2000);

/* If status equals TX_SUCCESS, my_pool is available
   for allocating memory. */
```

Figure 9.4: Creating a memory byte pool

```
TX_BYTE_POOL my_pool;
unsigned char *memory_ptr;
UINT status;

/* Allocate a 112 byte memory area from my_pool. Assume
   that the byte pool has already been created with a call
   to tx_byte_pool_create. */

status = tx_byte_allocate(&my_pool, (VOID **) &memory_ptr,
                          112, TX_WAIT_FOREVER);

/* If status equals TX_SUCCESS, memory_ptr contains the
   address of the allocated memory area. */
```

Figure 9.5: Allocating bytes from a memory byte pool

are allocated from the memory byte pool. To use this service, we must indicate how many bytes are needed, and what to do if enough memory is not available from this byte pool. Figure 9.5 shows a sample allocation, which will "wait forever" if adequate memory is not available. If the allocation succeeds, the pointer memory_ptr contains the starting location of the allocated bytes.

```
TX_BYTE_POOL my_pool;
UINT status;

...

/* Delete entire memory pool. Assume that the pool has already
   been created with a call to tx_byte_pool_create. */

status = tx_byte_pool_delete(&my_pool);

/* If status equals TX_SUCCESS, the memory pool is deleted. */
```

Figure 9.6: Deleting a memory byte pool

If variable *status* contains the return value TX_SUCCESS, then a block of 112 bytes, pointed to by memory_ptr has been created successfully.

Note that the time required by this service depends on the block size and the amount of fragmentation in the memory byte pool. Therefore, you should not use this service during time-critical threads of execution.

9.8 Deleting a Memory Byte Pool

A memory byte pool can be deleted with the tx_byte_pool_delete service. All threads that are suspended because they are waiting for memory from this byte pool are resumed and receive a TX_DELETED return status. Figure 9.6 shows how a memory byte pool can be deleted.

If variable *status* contains the return value TX_SUCCESS, then the memory byte pool has been deleted successfully.

9.9 Retrieving Memory Byte Pool Information

There are three services that enable you to retrieve vital information about memory byte pools. The first such service for memory byte pools—the tx_byte_pool_info_get service—retrieves a subset of information from the Memory Byte Pool Control Block. This information provides a "snapshot" at a particular instant in time, i.e., when the service is invoked. The other two services provide summary information that is based on the gathering of run-time performance data. One service—the tx_byte_pool_performance_info_get service—provides an information summary for a particular memory byte pool up to the time the service is invoked. By contrast the tx_byte_pool_performance_system_info_get

```
TX_BYTE_POOL my_pool;
CHAR *name;
ULONG available;
ULONG fragments;
TX_THREAD *first_suspended;
ULONG suspended_count;
TX_BYTE_POOL *next_pool;
UINT status;

...

/* Retrieve information about the previously created
   block pool "my_pool." */

status = tx_byte_pool_info_get(&my_pool, &name,
                               &available, &fragments,
                               &first_suspended, &suspended_count,
                               &next_pool);
/* If status equals TX_SUCCESS, the information requested is valid. */
```

Figure 9.7: Retrieving information about a memory byte pool

retrieves an information summary for all memory byte pools in the system up to the time the service is invoked. These services are useful in analyzing the behavior of the system and determining whether there are potential problem areas. The tx_byte_pool_info_get[4] service retrieves a variety of information about a memory byte pool. The information that is retrieved includes the byte pool name, the number of bytes available, the number of memory fragments, the location of the thread that is first on the suspension list for this byte pool, the number of threads currently suspended on this byte pool, and the location of the next created memory byte pool. Figure 9.7 shows how this service can be used to obtain information about a memory byte pool.

If variable *status* contains the return value TX_SUCCESS, then valid information about the memory byte pool has been obtained successfully.

9.10 Prioritizing a Memory Byte Pool Suspension List

When a thread is suspended because it is waiting for a memory byte pool, it is placed in the suspension list in a FIFO manner. When a memory byte pool regains an adequate

---

[4]By default, only the *tx_byte_pool_info_get* service is enabled. The other two information-gathering services must be enabled in order to use them.

```
TX_BYTE_POOL my_pool;
UINT status;

...

/* Ensure that the highest priority thread will receive
   the next free memory from this pool. */

status = tx_byte_pool_prioritize(&my_pool);

/* If status equals TX_SUCCESS, the highest priority
   suspended thread is at the front of the list. The
   next tx_byte_release call will wake up this thread,
   if there is enough memory to satisfy its request. */
```

Figure 9.8: Prioritizing the memory byte pool suspension list

amount of memory, the first thread in the suspension list (regardless of priority) receives an opportunity to allocate bytes from that memory byte pool. The tx_byte_pool_prioritize service places the highest-priority thread suspended for ownership of a specific memory byte pool at the front of the suspension list. All other threads remain in the same FIFO order in which they were suspended. Figure 9.8 shows how this service can be used.

If the variable *status* contains the value TX_SUCCESS, then the operation succeeded: the highest-priority thread in the suspension list has been placed at the front of the suspension list. The service also returns TX_SUCCESS if no thread was suspended on this memory byte pool. In this case the suspension list remains unchanged.

9.11 Releasing Memory to a Byte Pool

The tx_byte_release service releases a previously allocated memory area back to its associated pool. If one or more threads are suspended on this pool, each suspended thread receives the memory it requested and is resumed—until the pool's memory is exhausted or until there are no more suspended threads. This process of allocating memory to suspended threads always begins with the first thread on the suspension list. Figure 9.9 shows how this service can be used.

If the variable *status* contains the value TX_SUCCESS, then the memory block pointed to by memory_ptr has been returned to the memory byte pool.

```
unsigned char *memory_ptr;
UINT status;

…

/* Release a memory back to my_pool. Assume that the memory
   area was previously allocated from my_pool. */

status = tx_byte_release((VOID *) memory_ptr);

/* If status equals TX_SUCCESS, the memory pointed to by
   memory_ptr has been returned to the pool. */
```

Figure 9.9: Releasing bytes back to the memory byte pool

9.12 Memory Byte Pool Example—Allocating Thread Stacks

In the previous chapter, we used arrays to provide memory space for thread stacks. In this example, we will use a memory byte pool to provide memory space for the two threads. The first step is to declare the threads and a memory byte pool as follows:

```
TX_THREAD Speedy_Thread, Slow_Thread;
TX_MUTEX my_mutex;
#DEFINE STACK_SIZE 1024;
TX_BYTE_POOL my_pool;
```

Before we define the threads, we need to create the memory byte pool and allocate memory for the thread stack. Following is the definition of the byte pool, consisting of 4,500 bytes and starting at location 0×500000.

```
UINT status;
status = tx_byte_pool_create(&my_pool, "my_pool", (VOID *)
                             0×500000, 4500);
```

Assuming that the return value was TX_SUCCESS, we have successfully created a memory byte pool. Next, we allocate memory from this byte pool for the Speedy_Thread stack, as follows:

```
CHAR *stack_ptr;

status = tx_byte_allocate(&my_pool,
                          (VOID **) &stack_ptr,
                          STACK_SIZE, TX_WAIT_FOREVER);
```

Assuming that the return value was TX_SUCCESS, we have successfully allocated a block of memory for the stack, which is pointed to by stack_ptr. Next, we define Speedy_Thread using this block of memory for its stack (in place of the array stack_ speedy used in the previous chapter), as follows:

```
tx_thread_create(&Speedy_Thread, "Speedy_Thread",
                 Speedy_Thread_entry, 0,
                 stack_ptr, STACK_SIZE,
                 5, 5, TX_NO_TIME_SLICE, TX_AUTO_START);
```

We define the Slow_Thread in a similar fashion. The thread entry functions remain unchanged.

9.13 Memory Byte Pool Internals

When the TX_BYTE_POOL data type is used to declare a byte pool, a byte pool Control Block is created, and that Control Block is added to a doubly linked circular list, as illustrated in Figure 9.10.

The pointer named tx_byte_pool_created_ptr points to the first Control Block in the list. See the fields in the byte pool Control Block for byte pool attributes, values, and other pointers.

Allocations from memory byte pools resemble traditional malloc calls, which include the amount of memory desired (in bytes). ThreadX allocates from the pool in a first-fit manner, converts excess memory from this block into a new block, and places it back in the free memory list. This process is called *fragmentation*.

ThreadX *merges* free memory blocks together during a subsequent allocation search for a large enough free memory block. This process is called *defragmentation*.

The number of allocatable bytes in a memory byte pool is slightly less than what was specified during creation. This is because management of the free memory area introduces some overhead. Each free memory block in the pool requires the equivalent of two C pointers of overhead. In addition, when the pool is created ThreadX automatically allocates two blocks, a large free block and a small permanently allocated block at the end of the memory area. This allocated end block is used to improve performance of the allocation algorithm. It eliminates the need to continuously check for the end of the pool area during merging.

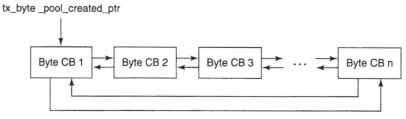

Figure 9.10: Created memory byte pool list

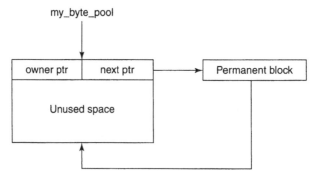

Figure 9.11: Organization of a memory byte pool upon creation

During run-time, the amount of overhead in the pool typically increases. This is partly because when an odd number of bytes is allocated, ThreadX pads out the allocated block to ensure proper alignment of the next memory block. In addition, overhead increases as the pool becomes more fragmented.

Figure 9.11 contains an illustration of a memory byte pool after it has been created, but before any memory allocations have occurred.

Initially, all usable memory space is organized into one contiguous block of bytes. However, each successive allocation from this byte pool can potentially subdivide the usable memory space. For example, Figure 9.12 shows a memory byte pool after the first memory allocation.

9.14 Summary of Memory Block Pools

Allocating memory in a fast and deterministic manner is essential in real-time applications. This is made possible by creating and managing multiple pools of fixed-size memory blocks called memory block pools.

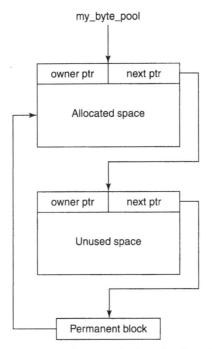

Figure 9.12: Memory byte pool after the first allocation

Because memory block pools consist of fixed-size blocks, using them involves no fragmentation problems. This is crucial because fragmentation causes behavior that is inherently nondeterministic. In addition, allocating and freeing fixed-size blocks is fast—the time required is comparable to that of simple linked-list manipulation. Furthermore, the allocation service does not have to search through a list of blocks when it allocates and deallocates from a memory block pool—it always allocates and deallocates at the head of the available list. This provides the fastest possible linked list processing and might help keep the currently used memory block in cache.

Lack of flexibility is the main drawback of fixed-size memory pools. The block size of a pool must be large enough to handle the worst-case memory requirements of its users. Making many different-sized memory requests from the same pool may cause memory waste. One possible solution is to create several different memory block pools that contain different sized memory blocks.

Each memory block pool is a public resource. ThreadX imposes no constraints as to how pools may be used. Applications may create memory block pools either during initialization or during run-time from within application threads. There is no limit to the number of memory block pools an application may use.

As noted earlier, memory block pools contain a number of fixed-size blocks. The block size, in bytes, is specified during creation of the pool. Each memory block in the pool imposes a small amount of overhead—the size of a C pointer. In addition, ThreadX may pad the block size in order to keep the beginning of each memory block on proper alignment.

The number of memory blocks in a pool depends on the block size and the total number of bytes in the memory area supplied during creation. To calculate the capacity of a pool (number of blocks that will be available), divide the block size (including padding and the pointer overhead bytes) into the total number of bytes in the supplied memory area.

The memory area for the block pool is specified during creation, and can be located anywhere in the target's address space. This is an important feature because of the considerable flexibility it gives the application. For example, suppose that a communication product has a high-speed memory area for I/O. You can easily manage this memory area by making it a memory block pool.

Application threads can suspend while waiting for a memory block from an empty pool. When a block is returned to the pool, ThreadX gives this block to the suspended thread and resumes the thread. If multiple threads are suspended on the same memory block pool, ThreadX resumes them in the order that they occur on the suspend thread list (usually FIFO).

However, an application can also cause the highest-priority thread to be resumed. To accomplish this, the application calls tx_block_pool_prioritize prior to the block release call that lifts thread suspension. The block pool prioritize service places the highest priority thread at the front of the suspension list, while leaving all other suspended threads in the same FIFO order.

9.15 Memory Block Pool Control Block

The characteristics of each memory block pool are found in its Control Block.[5] It contains information such as block size, and the number of memory blocks left. Memory

---

[5]The structure of the memory block pool Control Block is defined in the tx_api.h file.

Field	Description
tx_block_pool_id	Block pool ID
tx_block_pool_name	Pointer to block pool name
tx_block_pool_available	Number of available blocks
tx_block_pool_total	Initial number of blocks in the pool
tx_block_pool_available_list	Head pointer of the available block pool
tx_block_pool_start	Starting address of the block pool memory area
tx_block_pool_size	Block pool size in bytes
tx_block_pool_block_size	Individual memory block size - rounded
*tx_block_pool_suspension_list	Block pool suspension list head
tx_block_pool_suspended_count	Number of threads suspended
*tx_block_pool_created_next	Pointer to the next block pool in the created list
*tx_block_pool_created_previous	Pointer to the previous block pool in the created list

Figure 9.13: Memory Block Pool Control Block

pool Control Blocks can be located anywhere in memory, but they are commonly defined as global structures outside the scope of any function. Figure 9.13 lists most members of the memory pool Control Block.

The user of an allocated memory block must not write outside its boundaries. If this happens, corruption occurs in an adjacent (usually subsequent) memory area. The results are unpredictable and quite often catastrophic.

In most cases, the developer can ignore the contents of the Memory Block Pool Control Block. However, there are several fields that may be useful during debugging, such as the number of available blocks, the initial number of blocks, the actual block size, the total number of bytes in the block pool, and the number of threads suspended on this memory block pool.

9.16 Summary of Memory Block Pool Services

Appendix A contains detailed information about memory block pool services. This appendix contains information about each service, such as the prototype, a brief description of the service, required parameters, return values, notes and warnings, allowable invocation, and an example showing how the service can be used. Figure 9.14 contains a list of all available memory block pool services. In the succeeding sections of this chapter, we will investigate each of these services.

Memory block pool service	Description
tx_block_allocate	Allocate a fixed-size block of memory
tx_block_pool_create	Create a pool of fixed-size memory blocks
tx_block_pool_delete	Delete a memory block pool
tx_block_pool_info_get	Retrieve information about a memory block pool
tx_block_pool_performance_info_get	Get block pool performance information
tx_block_pool_performance_system_info_get	Get block pool system performance information
tx_block_pool_prioritize	Prioritize the memory block pool suspension list
tx_block_release	Release a fixed-sized memory block

Figure 9.14: Services of the memory block pool

We will first consider the tx_block_pool_create service because it must be invoked before any of the other services.

9.17 Creating a Memory Block Pool

A memory block pool is declared with the TX_BLOCK_POOL data type and is defined with the tx_block_pool_create service. When defining a memory block pool, you need to specify its Control Block, the name of the memory block pool, the address of the memory block pool, and the number of bytes available. Figure 9.15 contains a list of these attributes.

We will develop one example of memory block pool creation to illustrate the use of this service, and we will name it "my_pool." Figure 9.16 contains an example of memory block pool creation.

If variable *status* contains the return value TX_SUCCESS, then we have successfully created a memory block pool called my_pool that contains a total of 1,000 bytes, with each block containing 50 bytes. The number of blocks can be calculated as follows:

$$Total\ Number\ of\ Blocks = \frac{Total\ number\ of\ Bytes\ Available}{(Number\ of\ Bytes\ in\ Each\ Memory\ Block) + (size\ of\ (void*))}$$

Memory block pool control block
Name of memory block pool
Number of bytes in each fixed-size memory block
Starting address of the memory block pool
Total number of bytes available to the block pool

Figure 9.15: Memory block pool attributes

```
TX_BLOCK_POOL my_pool;
UINT status;

/* Create a memory pool whose total size is 1000 bytes
   starting at address 0x100000. Each block in this
   pool is defined to be 50 bytes long. */

status = tx_block_pool_create(&my_pool, "my_pool_name",
                        50, (VOID *) 0x100000, 1000);

/* If status equals TX_SUCCESS, my_pool contains about 18
   memory blocks of 50 bytes each. The reason there are
   not 20 blocks in the pool is because of the one
   overhead pointer associated with each block. */
```

Figure 9.16: Creating a memory block pool

Assuming that the value of size of (*void**) is four bytes, the total number of blocks available is calculated thus:

$$Total\ Number\ of\ Blocks = \frac{1000}{(50)+(4)} = 18.52 = 18\ blocks$$

Use the preceding formula to avoid wasting space in a memory block pool. Be sure to carefully estimate the needed block size and the amount of memory available to the pool.

9.18 Allocating a Memory Block Pool

After a memory block pool has been declared and defined, we can start using it in a variety of applications. The tx_block_allocate service is the method that allocates a fixed size block of memory from the memory block pool. Because the size of the memory block pool is determined when it is created, we need to indicate what to do if enough memory is not available from this block pool. Figure 9.17 contains an example of allocating one block from a memory block pool, in which we will "wait forever" if adequate memory is not available. After memory allocation succeeds, the pointer memory_ptr contains the starting location of the allocated fixed-size block of memory.

If variable *status* contains the return value TX_SUCCESS, then we have successfully allocated one fixed-size block of memory. This block is pointed to by memory_ptr.

9.19 Deleting a Memory Block Pool

A memory block pool can be deleted with the tx_block_pool_delete service. All threads that are suspended because they are waiting for memory from this block pool are resumed and receive a TX_DELETED return status. Figure 9.18 shows how a memory block pool can be deleted.

```
TX_BLOCK_POOL my_pool;
unsigned char *memory_ptr;
UINT status;

...

/* Allocate a memory block from my_pool. Assume that the
   pool has already been created with a call to
   tx_block_pool_create. */

status = tx_block_allocate(&my_pool, (VOID **) &memory_ptr,
                     TX_WAIT_FOREVER);

/* If status equals TX_SUCCESS, memory_ptr contains the
   address of the allocated block of memory. */
```

Figure 9.17: Allocation of a fixed-size block of memory

```
TX_BLOCK_POOL my_pool;
UINT status;
...

/* Delete entire memory block pool. Assume that the
   pool has already been created with a call to
   tx_block_pool_create. */

status = tx_block_pool_delete(&my_pool);

/* If status equals TX_SUCCESS, the memory block pool
   has been deleted. */
```

Figure 9.18: Deleting a memory block pool

If variable *status* contains the return value TX_SUCCESS, then we have successfully deleted the memory block pool.

9.20 Retrieving Memory Block Pool Information

The tx_block_pool_info_get service retrieves a variety of information about a memory block pool. The information that is retrieved includes the block pool name, the number of blocks available, the total number of blocks in the pool, the location of the thread that is first on the suspension list for this block pool, the number of threads currently suspended on this block pool, and the location of the next created memory block pool. Figure 9.19 show how this service can be used to obtain information about a memory block pool.

If variable *status* contains the return value TX_SUCCESS, then we have successfully obtained valid information about the memory block pool.

9.21 Prioritizing a Memory Block Pool Suspension List

When a thread is suspended because it is waiting for a memory block pool, it is placed in the suspension list in a FIFO manner. When a memory block pool regains a block

```
TX_BLOCK_POOL my_pool;
CHAR *name;
ULONG available;
ULONG total_blocks;
TX_THREAD *first_suspended;
ULONG suspended_count;
TX_BLOCK_POOL *next_pool;
UINT status;

/* Retrieve information about the previously created
   block pool "my_pool." */

status = tx_block_pool_info_get(&my_pool, &name,
                                &available,&total_blocks,
                                &first_suspended,
                                &suspended_count,
                                &next_pool);

/* If status equals TX_SUCCESS, the information requested
   is valid. */
```

Figure 9.19: Retrieving information about a memory block pool

```
TX_BLOCK_POOL my_pool;
UINT status;

…

/* Ensure that the highest priority thread will receive
   the next free block in this pool. */

status = tx_block_pool_prioritize(&my_pool);

/* If status equals TX_SUCCESS, the highest priority
   suspended thread is at the front of the list. The
   next tx_block_release call will wake up this thread. */
```

Figure 9.20: Prioritizing the memory block pool suspension list

of memory, the first thread in the suspension list (regardless of priority) receives an opportunity to take a block from that memory block pool. The tx_block_pool_prioritize service places the highest-priority thread suspended for ownership of a specific memory block pool at the front of the suspension list. All other threads remain in the same FIFO order in which they were suspended. Figure 9.20 contains an example showing how this service can be used.

If the variable *status* contains the value TX_SUCCESS, the prioritization request succeeded. The highest-priority thread in the suspension list that is waiting for the memory block pool called "my_pool" has moved to the front of the suspension list. The service call also returns TX_SUCCESS if no thread was waiting for this memory block pool. In this case, the suspension list remains unchanged.

9.22 Releasing a Memory Block

The tx_block_release service releases one previously allocated memory block back to its associated block pool. If one or more threads are suspended on this pool, each suspended thread receives a memory block and is resumed until the pool runs out of blocks or until there are no more suspended threads. This process of allocating memory to suspended threads always begins with the first thread on the suspended list. Figure 9.21 shows how this service can be used.

If the variable *status* contains the value TX_SUCCESS, then the memory block pointed to by memory_ptr has been returned to the memory block pool.

```
TX_BLOCK_POOL my_pool;
unsigned char *memory_ptr;
UINT status;
…

/* Release a memory block back to my_pool. Assume that the
   pool has been created and the memory block has been
   allocated. */

status = tx_block_release((VOID *) memory_ptr);

/* If status equals TX_SUCCESS, the block of memory pointed
   to by memory_ptr has been returned to the pool. */
```

Figure 9.21: Release one block to the memory block pool

9.23 Memory Block Pool Example—Allocating Thread Stacks

In the previous chapter, we allocated thread stack memory from arrays, and earlier in this chapter we allocated thread stacks from a byte pool. In this example, we will use a memory block pool. The first step is to declare the threads and a memory block pool as follows:

```
TX_THREAD Speedy_Thread, Slow_Thread;
TX_MUTEX my_mutex;
#DEFINE STACK_SIZE 1024;
TX_BLOCK_POOL my_pool;
```

Before we define the threads, we need to create the memory block pool and allocate memory for the thread stack. Following is the definition of the block pool, consisting of four blocks of 1,024 bytes each and starting at location 0×500000.

```
UINT status;
status = tx_block_pool_create(&my_pool, "my_pool", 1024,
                              (VOID *) 0×500000, 4520);
```

Assuming that the return value was TX_SUCCESS, we have successfully created a memory block pool. Next, we allocate memory from that block pool for the Speedy_ Thread stack, as follows:

```
CHAR *stack_ptr;
```

```
status = tx_block_allocate(&my_pool,
                (VOID **) &stack_ptr,
                TX_WAIT_FOREVER);
```

Assuming that the return value was TX_SUCCESS, we have successfully allocated a block of memory for the stack, which is pointed to by stack_ptr. Next, we define Speedy_Thread by using that block of memory for its stack, as follows:

```
tx_thread_create(&Speedy_Thread, "Speedy_Thread",
                Speedy_Thread_entry, 0,
                stack_ptr, STACK_SIZE,
                5, 5, TX_NO_TIME_SLICE, TX_AUTO_START);
```

We define the Slow_Thread in a similar fashion. The thread entry functions remain unchanged.

9.24 Memory Block Pool Internals

When the TX_BLOCK_POOL data type is used to declare a block pool, a Block Pool Control Block is created, and that Control Block is added to a doubly linked circular list, as illustrated in Figure 9.22.

The pointer named tx_block_pool_created_ptr points to the first Control Block in the list. See the fields in the Block Pool Control Block for block pool attributes, values, and other pointers.

As noted earlier, block pools contain fixed-size blocks of memory. The advantages of this approach include fast allocation and release of blocks, and no fragmentation issues. One possible disadvantage is that space could be wasted if the block size is too large. However, developers can minimize this potential problem by creating several block pools

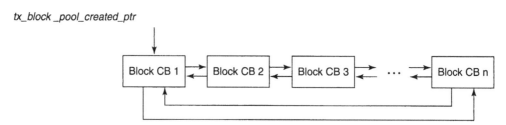

Figure 9.22: Created memory block pool list

with different block sizes. Each block in the pool entails a small amount of overhead, i.e., an owner pointer and a next block pointer. Figure 9.23 illustrates the organization of a memory block pool.

9.25 Overview and Comparison

We considered two approaches for memory management in this chapter. The first approach is the memory byte pool, which allows groups of bytes of variable size to be used and reused. This approach has the advantage of simplicity and flexibility, but leads

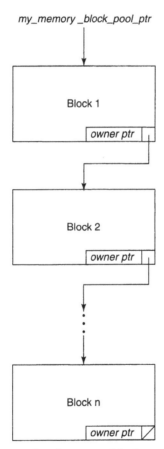

Figure 9.23: Example of memory block pool organization

to fragmentation problems. Because of fragmentation, a memory byte pool may have enough total bytes to satisfy a request for memory, but can fail to satisfy that request because the available bytes are not contiguous. Therefore, we generally recommend that you avoid using memory byte pools for deterministic, real-time applications.

The second approach for memory management is the memory block pool, which consists of fixed-size memory blocks, thus eliminating the fragmentation problem. Memory block pools lose some flexibility because all memory blocks are the same size, and a given application may not need that much space. However, developers can alleviate this problem by creating several memory block pools, each with a different block size. Furthermore, allocating and releasing memory blocks is fast and predictable. In general, we recommend the use of memory block pools for deterministic, real-time applications.

9.26 Key Terms and Phrases

allocation of memory	fixed-size block
block size calculation	fragmentation
creating memory pools	information retrieval
defragmentation	memory block pool
deleting memory pools	Memory Block Pool Control Block
memory byte pool	release of memory
Memory Byte Pool Control Block	suspend while waiting for memory
pointer overhead	thread stack memory allocation
prioritize memory pool suspension list	wait option

9.27 Problems

1. Memory block pools are recommended for deterministic, real-time applications. Under what circumstances should you use memory byte pools?

2. In the previous chapter, thread stacks were allocated from arrays. What advantages do memory block pools have over arrays when providing memory for thread stacks?

3. Suppose that an application has created a memory byte pool and has made several allocations from it. The application then requests 200 bytes when the pool has 500 total bytes available. Explain why the pool might not fulfill that request in a timely fashion.

4. Suppose that a memory block pool is created with a total of 1,000 bytes, with each block 100 bytes in size. Explain why this pool contains fewer than 10 blocks.

5. Create the memory block pool in the previous problem and inspect the following fields in the Control Block: tx_block_pool_available, tx_block_pool_size, tx_block_pool_block_size, and tx_block_pool_suspended_count.

6. The section titled *Memory Block Pool Example—Allocating Thread Stacks* contains a definition for Speedy_Thread using a block of memory for its stack. Develop a definition for Slow_Thread using a block of memory for its stack, and then compile and execute the resulting system.

CHAPTER 10

Internal System Clock and Application Timers

10.1 Introduction

Fast response to asynchronous external events is the most important function of real-time, embedded applications. However, many of these applications must also perform certain activities at predetermined intervals of time.

ThreadX application timers enable you to execute application C functions at specific intervals of time. You can also set an application timer to expire only once. This type of timer is called a *one-shot timer*, while repeating interval timers are called *periodic timers*. Each application timer is a public resource.

Time intervals are measured by periodic timer interrupts. Each timer interrupt is called a timer-tick. The actual time between timer-ticks is specified by the application, but 10 ms is the norm for many implementations.[1]

The underlying hardware must be able to generate periodic interrupts in order for application timers to function. In some cases, the processor has a built-in periodic interrupt capability. If not, the user's computer board must have a peripheral device that can generate periodic interrupts.

ThreadX can still function even without a periodic interrupt source. However, all timer-related processing is then disabled. This includes time-slicing, suspension timeouts, and timer services.

---

[1]The periodic timer setup is typically found in the *tx_Initialize_low_level* assembly file.

Timer expiration intervals are specified in terms of timer-ticks. The timer count starts at the specified expiration value and decreases by one on each timer-tick. Because an application timer could be enabled just prior to a timer interrupt (or timer-tick), the timer could expire up to one tick early.

If the timer-tick rate is 10 ms, application timers may expire up to 10 ms early. This inaccuracy is more significant for 10 ms timers than for one-second timers. Of course, increasing the timer interrupt frequency decreases this margin of error.

Application timers execute in the order in which they become active. For example, if you create three timers with the same expiration value and then activate them, their corresponding expiration functions[2] are guaranteed to execute in the order that you activated the timers.

By default, application timers execute from within a hidden system thread that runs at priority zero, which is higher than any application thread. Therefore, you should keep processing inside timer expiration functions to a minimum. It is important to avoid suspending on any service calls made from within the application timer's expiration function.

It is also important to avoid, whenever possible, using timers that expire every timer tick. This might induce excessive overhead in the application.

In addition to the application timers, ThreadX provides a single continuously incrementing 32-bit tick counter. This tick counter, or internal system clock, increments by one on each timer interrupt. An application can read or set this 32-bit counter with calls to tx_time_get and tx_time_set, respectively. The use of the internal system clock is determined completely by the application.

We will first consider the two services for the internal system clock (i.e., tx_time_get and tx_time_set), and then we will investigate application timers.

10.2 Internal System Clock Services

ThreadX sets the internal system clock to zero during application initialization, and each timer-tick[3] increases the clock by one. The internal system clock is intended for the sole

---

[2]An expiration function is sometimes called a timeout function.

[3]The actual time represented by each timer-tick is application-specific.

```
ULONG current_time;

/* Retrieve the current system time, in timer-ticks. */

current_time = tx_time_get();

/* Variable current_time now contains the current system time */
```

Figure 10.1: Get current time from the internal system clock

```
/* Set the internal system time to 0x1234. */

tx_time_set(0x1234);

/* Current time now contains 0x1234 until the next
   timer interrupt. */
```

Figure 10.2: Set current time of the internal system clock

use of the developer; ThreadX does not use it for any purpose, including implementing application timers. Applications can perform exactly two actions on the internal system clock: either read the current clock value, or set its value. Appendix I contains additional information about internal system clock services.

The tx_time_get service retrieves the current time from the internal system clock. Figure 10.1 illustrates how this service can be used.

After invoking this service, the variable current_time contains a copy of the internal system clock. This service can be used to measure elapsed time and perform other time-related calculations.

The tx_time_set service sets the current time of the internal system clock to some specified value. Figure 10.2 illustrates how this service can be used.

After invoking this service, the current time of the internal system clock contains the value 0×1234. The time will remain at this value until the next timer-tick, when it will be incremented by one.

---

[4]The structure of the Application Timer Control Block is defined in the tx_api.h file.

Field	Description
tx_timer_id	Application timer ID
tx_timer_name	Pointer to application timer name
tx_remaining_ticks	Number of remaining timer-ticks
tx_re_initialize_ticks	Re-initialization timer-tick value
tx_timeout_function	Pointer to timeout function
tx_timeout_param	Parameter for timeout function
tx_active_next	Pointer to next active internal timer
tx_active_previous	Pointer to previous active internal timer
tx_list_head	Pointer to head of list of internal timers
tx_timer_created_next	Pointer to the next timer in the created list
tx_timer_created_previous	Pointer to the previous timer in the created list

Figure 10.3: Application Timer Control Block

10.3 Application Timer Control Block

The characteristics of each application timer are found in its Application Timer Control Block.[4] It contains useful information such as the ID, application timer name, the number of remaining timer-ticks, the re-initialization value, the pointer to the timeout function, the parameter for the timeout function, and various pointers. As with the other Control Blocks, ThreadX prohibits an application from explicitly modifying the Application Timer Control Block.

Application Timer Control Blocks can be located anywhere in memory, but it is most common to make the Control Block a global structure by defining it outside the scope of any function. Figure 10.3 contains many of the fields that comprise this Control Block.

An Application Timer Control Block is created when an application timer is declared with the TX_TIMER data type. For example, we declare my_timer as follows:

```
TX_TIMER my_timer;
```

The declaration of application timers normally appears in the declaration and definition section of the application program.

10.4 Summary of Application Timer Services

Appendix J contains detailed information about application timer services. This appendix contains information about each service, such as the prototype, a brief description of the service, required parameters, return values, notes and warnings, allowable invocation,

Application timer service	Description
tx_timer_activate	Activate an application timer
tx_timer_change	Change characteristics of an application timer
tx_timer_create	Create an application timer
tx_timer_deactivate	Deactivate an application timer
tx_timer_delete	Delete an application timer
tx_timer_info_get	Retrieve information about an application timer
tx_timer_performance_info_get	Get timer performance information
tx_timer_performance_system_info_get	Get timer system performance information

Figure 10.4: Services of the application timer

and an example showing how the service can be used. Figure 10.4 contains a listing of all available application timer services. We will investigate each of these services in the subsequent sections of this chapter.

We will first consider the tx_timer_pool_create service because it needs to be invoked before any of the other services.

10.5 Creating an Application Timer

An application timer is declared with the TX_TIMER data type and is defined with the tx_timer_create service. When defining an application timer, you must specify its Control Block, the name of the application timer, the expiration function to call when the timer expires, the input value to pass to the expiration function, the initial number of timer-ticks before timer expiration, the number of timer-ticks for all timer expirations after the first, and the option that determines when the timer is activated. The valid range of values for the initial number of timer-ticks is from 1 to 0xFFFFFFFF (inclusive).

For subsequent time timer-ticks, the valid range of values is from 0 to 0xFFFFFFFF (inclusive), where the value of 0 means this is a *one-shot timer*, and all other values in that range are for *periodic timers*. Figure 10.5 contains a list of these attributes.

We will illustrate the application timer creation service with an example. We will give our application timer the name "my_timer" and cause it to activate immediately. The timer

Application timer control block
Application timer name
Expiration function to call when the timer expires
Input value to pass to the expiration function
Initial number of timer-ticks
Number of timer-ticks for all timer expirations after the first
Auto activate option

Figure 10.5: Attributes of an application timer

```
TX_TIMER my_timer;
UINT status;

/* Create an application timer that executes
   "my_timer_function" after 100 timer-ticks initially and then
   after every 25 timer-ticks. This timer is specified to start
   immediately. */

status = tx_timer_create(&my_timer,"my_timer_name",
                         my_timer_function, 0x1234, 100, 25,
                         TX_AUTO_ACTIVATE);

/* If status equals TX_SUCCESS, my_timer_function will
   be called 100 timer-ticks later and then called every
   25 timer-ticks. Note that the value 0x1234 is passed to
   my_timer_function every time it is called. */
```

Figure 10.6: Creating an application timer

will expire after 100 timer-ticks, call the expiration function called "my_timer_function," and will continue to do so every 25 timer-ticks thereafter. Figure 10.6 contains an example of application timer creation.

If variable *status* contains the return value TX_SUCCESS, we have successfully created the application timer. We must place a prototype for the expiration function in the declaration and definition section of our program as follows:

```
void my_timer_function (ULONG);
```

The expiration function definition appears in the final section of the program, where the thread entry functions are defined. Following is a skeleton of that function definition.

```
        TX_TIMER my_timer;
        UINT status;

        ...

        /* Activate an application timer. Assume that the
           application timer has already been created. */

        status = tx_timer_activate(&my_timer);

        /* If status equals TX_SUCCESS, the application timer is
           now active. */
```

Figure 10.7: Activation of an application timer

```
void my_timer_function (ULONG invalue)
{
  :
}
```

10.6 Activating an Application Timer

When an application timer is created with the TX_NO_ACTIVATE option, it remains inactive until the tx_timer_activate service is called. Similarly, if an application timer is deactivated with the tx_timer_deactive service, it remains inactive until the tx_timer_ activate service is called. If two or more application timers expire at the same time, the corresponding expiration functions are executed in the order in which they were activated. Figure 10.7 contains an example of application timer activation.

10.7 Changing an Application Timer

When you create an application timer, you must specify the initial number of timer-ticks before timer expiration, as well as the number of timer-ticks for all timer expirations after the first. Invoking the tx_timer_change service can change these values. You must deactivate the application timer before calling this service, and call tx_timer_activate after this service to restart the timer. Figure 10.8 illustrates how this service can be called.

If variable *status* contains the return value TX_SUCCESS, we have successfully changed the number of timer-ticks for initial and subsequent expiration to 50.

```
TX_TIMER my_timer;
UINT status;

…

/* Change a previously created and now deactivated timer
   to expire every 50 timer-ticks, including the initial
   expiration. */

status = tx_timer_change(&my_timer,50, 50);

/* If status equals TX_SUCCESS, the specified timer is
   changed to expire every 50 timer-ticks. */

/* Activate the specified timer to get it started again. */
   status = tx_timer_activate(&my_timer);
```

Figure 10.8: Change characteristics of an application timer

```
TX_TIMER my_timer;
UINT status;

…

/* Deactivate an application timer. Assume that the
   application timer has already been created. */

status = tx_timer_deactivate(&my_timer);

/* If status equals TX_SUCCESS, the application timer
   is now deactivated. */
```

Figure 10.9: Deactivate an application timer

10.8 Deactivating an Application Timer

Before modifying the timing characteristics of an application timer, that timer must first be deactivated. This is the sole purpose of the tx_timer_deactivate service. Figure 10.9 shows how to use this service.

If variable *status* contains the value TX_SUCCESS, the application timer is now deactivated. This timer remains in an inactive state until it is activated with the tx_timer_activate service.

```
TX_TIMER my_timer;
UINT status;

…

/* Delete application timer. Assume that the
   application timer has already been created. */

status = tx_timer_delete(&my_timer);

/* If status equals TX_SUCCESS, the application
   timer is deleted. */
```

Figure 10.10: Deleting an application timer

10.9 Deleting an Application Timer

The tx_timer_delete service deletes an application timer. Figure 10.10 shows how to delete an application timer.

If variable *status* contains the return value TX_SUCCESS, we have successfully deleted the application timer. Make certain that you do not inadvertently use a deleted timer.

10.10 Retrieving Application Timer Information

There are three services that enable you to retrieve vital information about application timers. The first such service for application timers—the tx_timer_pool_info_get service—retrieves a subset of information from the Application Timer Control Block. This information provides a "snapshot" at a particular instant in time, i.e., when the service is invoked. The other two services provide summary information that is based on the gathering of run-time performance data. One service—the tx_timer_pool_performance_info_get service—provides an information summary for a particular application timer up to the time the service is invoked. By contrast the tx_timer_pool_performance_system_info_get retrieves an information summary for all application timers in the system up to the time the service is invoked. These services are useful in analyzing the behavior of the system and determining whether there are potential problem areas. The tx_timer_info_get[5] service retrieves a variety of information about an

---

[5]By default, only the *tx_timer_info_get* service is enabled. The other two information-gathering services must be enabled in order to use them.

```
TX_TIMER my_timer;
CHAR *my_timer_name;
UINT active;
ULONG remaining_ticks;
ULONG reschedule_ticks;
TX_TIMER *next_timer;
UINT status;

...

/* Retrieve information about the previously created
   application timer called "my_timer." */

status = tx_timer_info_get(&my_timer, &my_timer_name,
                           &active,&remaining_ticks,
                           &reschedule_ticks,
                           &next_timer);
/* If status equals TX_SUCCESS, the information
   requested is valid. */
```

Figure 10.11: Retrieve information about an application timer

application timer. The information that is retrieved includes the application timer name, its active/inactive state, the number of timer-ticks before the timer expires, the number of subsequent timer-ticks for timer expiration after the first expiration, and a pointer to the next created application timer. Figure 10.11 shows how this service can be used to obtain information about an application timer.

If variable *status* contains the return value TX_SUCCESS, we have retrieved valid information about the timer.

10.11 Sample System Using Timers to Measure Thread Performance

In Chapter 8, we created a sample system that produced output beginning as follows:

```
Current Time: 34 Speedy_Thread finished cycle...
Current Time: 40 Slow_Thread finished cycle...
Current Time: 56 Speedy_Thread finished cycle...
```

The timing data in this output was captured with the use of the internal system clock. Following are the statements near the end of the Speedy_Thread entry function that generate information about the performance of Speedy_Thread.

```
/* Declare the application timer */
TX_TIMER        stats_timer;

/* Declare the counters and accumulators */
ULONG           Speedy_Thread_counter = 0,
                total_speedy_time = 0;
ULONG           Slow_Thread_counter = 0,
                total_slow_time = 0;

/* Define prototype for expiration function */
void    print_stats(ULONG);
```

Figure 10.12: Additions to the declarations and definitions section

```
current_time = tx_time_get();
printf("Current Time: %5lu Speedy_Thread finished cycle...\n",
current_time);
```

We used the tx_time_get service to retrieve the current time from the internal system clock. Each time the Speedy_Thread finished its cycle it displayed the preceding timing information.

For this sample system, we will eliminate the display of information at the end of each thread cycle. However, we will continue to use the internal system clock to determine the time duration of a thread cycle, but we will display a summary of information at designated intervals, say every 500 timer-ticks. We will use an application timer to trigger this periodic display. Following is a portion of the sample output we would like to have displayed on a periodic basis:

```
**** Timing Info Summary
    Current Time:           500
Speedy_Thread counter:      22
    Speedy_Thread avg time: 22
    Slow_Thread counter:    11
    Slow_Thread avg time:   42
```

We need to compute the average cycle time for both the Speedy_Thread and the Slow_Thread. To accomplish this, we need two variables for each of the two threads: one to store the total time spent in the cycle, and one to count the total number of cycles completed. Figure 10.12 contains these variable declarations and the other additions to the declarations and definitions section of the program.

```
/* Create and activate the timer */
tx_timer_create (&stats_timer, "stats_timer", print_stats,
                0x1234, 500, 500, TX_AUTO_ACTIVATE);
```

Figure 10.13: Additions to the application definitions section

```
/* Insert at the beginning of Speedy_Thread entry function */
   ULONG    start_time, cycle_time;

/* Get the starting time for this cycle */
   start_time = tx_time_get();

...


/* Insert at the end of Speedy_Thread entry function */
/* Increment thread counter, compute cycle time & total time */
   Speedy_Thread_counter++;
   current_time = tx_time_get();
   cycle_time = current_time - start_time;
   total_speedy_time = total_speedy_time + cycle_time;
```

Figure 10.14: Additions to the Speedy_Thread entry function

We need to add the timer creation service to the tx_application_define function, as indicated by Figure 10.13.

We need to modify the entry functions for the Speedy_Thread and the Slow_Thread. Delete the following statements at the end of the Speedy_Thread entry function:

```
current_time = tx_time_get();
printf("Current Time: %lu Speedy_Thread finished cycle...\n",
current_time);
```

We will use the internal system clock to compute the time to complete each cycle. We also need to compute the total cycle time as follows:

$$total\_speedy\_time = \Sigma cycle\_time$$

Figure 10.14 contains the necessary additions for the Speedy_Thread.

The entry function for Slow_Thread requires similar additions, but we leave that as an exercise for the reader. These computations store the total number of cycles that have

```
/* Display statistics at periodic intervals */
void print_stats (ULONG invalue)
{
    ULONG    current_time, avg_slow_time, avg_speedy_time;
    if ((Speedy_Thread_counter>0) && (Slow_Thread_counter>0))
    {
        current_time = tx_time_get();
        avg_slow_time = total_slow_time / Slow_Thread_counter;
        avg_speedy_time = total_speedy_time / Speedy_Thread_counter;

        printf("\n**** Timing Info Summary\n\n");
        printf("Current Time:                %lu\n", current_time);
        printf("  Speedy_Thread counter:     %lu\n", Speedy_Thread_counter);
        printf(" Speedy_Thread avg time:     %lu\n", avg_speedy_time);
        printf("    Slow_Thread counter:     %lu\n", Slow_Thread_counter);
        printf("   Slow_Thread avg time:     %lu\n\n", avg_slow_time);
    }
    else printf("Bypassing print_stats, Time: %lu\n", tx_time_get());
}
```

Figure 10.15: Expiration function to display summary information

been completed, and the total amount of time spent in those cycles. The expiration function called print_stats will use these values to compute average cycle time for both threads and will display summary information.

Every 500 timer-ticks, the application timer called stats_timer expires and invokes the expiration function print_stats. After determining that both thread counters are greater than zero, that function computes the average cycle times for Speedy_Thread and Slow_Thread, as follows:

$$avg\_speedy\_time = \frac{total\_speedy\_time}{speedy\_thread\_counter}$$

and

$$avg\_slow\_time = \frac{total\_slow\_time}{slow\_thread\_counter}$$

Function print_stats then displays the current time, the average cycle times, and the number of cycles completed by each of the threads. Figure 10.15 contains a listing of the print_stats expiration function.

This program can be modified easily to display other timing information; the timer can be changed to expire at different time intervals as well. The complete program listing called *10_sample_system.c* is located in the next section of this chapter and on the attached CD.

10.12 Listing for 10_sample_system.c

The sample system named *10_sample_system.c* is located on the attached CD. The complete listing appears below; line numbers have been added for easy reference.

```
001  /* 10_sample_system.c
002
003  Create two threads, and one mutex.
004  Use arrays for the thread stacks.
005  The mutex protects the critical sections.
006  Use an application timer to display thread timings. */
007
008  /****************************************************/
009  /*      Declarations, Definitions, and Prototypes    */
010  /****************************************************/
011
012  #include "tx_api.h"
013  #include <stdio.h >
014
015  #define   STACK_SIZE       1024
016
017  CHAR stack_speedy[STACK_SIZE];
018  CHAR stack_slow[STACK_SIZE];
019
020
021  /* Define the ThreadX object control blocks... */
022
023  TX_THREAD     Speedy_Thread;
024  TX_THREAD     Slow_Thread;
025
026  TX_MUTEX      my_mutex;
027
028  /* Declare the application timer */
029  TX_TIMER stats_timer;
030
```

```
031   /* Declare the counters and accumulators */
032   ULONG        Speedy_Thread_counter = 0,
033                total_speedy_time = 0;
034   ULONG        Slow_Thread_counter = 0,
035                total_slow_time = 0;
036
037   /* Define prototype for expiration function */
038   void print_stats(ULONG);
039        .
040   /* Define thread prototypes. */
041
042   void Speedy_Thread_entry(ULONG thread_input);
043   void Slow_Thread_entry(ULONG thread_input);
044
045
046   /***************************************************/
047   /*                Main Entry Point                 */
048   /***************************************************/
049
050   /* Define main entry point. */
051
052   int main()
053   {
054
055       /* Enter the ThreadX kernel. */
056       tx_kernel_enter();
057   }
058
059
060   /***************************************************/
061   /*             Application Definitions             */
062   /***************************************************/
063
064   /* Define what the initial system looks like. */
065
066   void tx_application_define(void *first_unused_memory)
067   {
068
069
070       /* Put system definitions here,
```

```
071              e.g., thread and mutex creates */
072
073        /* Create the Speedy_Thread. */
074        tx_thread_create(&Speedy_Thread, "Speedy_Thread",
075                         Speedy_Thread_entry, 0,
076                         stack_speedy, STACK_SIZE,
077                         5, 5, TX_NO_TIME_SLICE, TX_AUTO_START);
078
079        /* Create the Slow_Thread */
080        tx_thread_create(&Slow_Thread, "Slow_Thread",
081                         Slow_Thread_entry, 1,
082                         stack_slow, STACK_SIZE,
083                         15, 15, TX_NO_TIME_SLICE, TX_AUTO_START);
084
085        /* Create the mutex used by both threads */
086        tx_mutex_create(&my_mutex, "my_mutex", TX_NO_INHERIT);
087
088  /* Create and activate the timer */
089        tx_timer_create (&stats_timer, "stats_timer", print_stats,
090                         0X1234, 500, 500, TX_AUTO_ACTIVATE);
091
092  }
093
094
095  /****************************************************/
096  /*                Function Definitions              */
097  /****************************************************/
098
099  /* Define the activities for the Speedy_Thread */
100
101  void Speedy_Thread_entry(ULONG thread_input)
102  {
103  UINT  status;
104  ULONG current_time;
105  ULONG start_time, cycle_time;
106
107      while(1)
108      {
109
110          /* Get the starting time for this cycle */
```

```
111          start_time = tx_time_get();
112
113          /* Activity 1: 2 timer-ticks. */
114          tx_thread_sleep(2);
115
116          /* Activity 2: 5 timer-ticks *** critical section ***
117          Get the mutex with suspension. */
118
119          status = tx_mutex_get(&my_mutex, TX_WAIT_FOREVER);
120          if (status != TX_SUCCESS) break; /* Check status */
121
122          tx_thread_sleep(5);
123
124          /* Release the mutex. */
125          status = tx_mutex_put(&my_mutex);
126          if (status != TX_SUCCESS) break; /* Check status */
127
128          /* Activity 3: 4 timer-ticks. */
129          tx_thread_sleep(4);
130
131          /* Activity 4: 3 timer-ticks *** critical section ***
132          Get the mutex with suspension. */
133
134          status = tx_mutex_get(&my_mutex, TX_WAIT_FOREVER);
135          if (status != TX_SUCCESS) break; /* Check status */
136
137          tx_thread_sleep(3);
138
139          /* Release the mutex. */
140          status = tx_mutex_put(&my_mutex);
141          if (status != TX_SUCCESS) break; /* Check status */
142
143          /* Increment thread counter, compute cycle time & total
                time */
144          Speedy_Thread_counter++;
145          current_time = tx_time_get();
146          cycle_time = current_time-start_time;
147          total_speedy_time = total_speedy_time + cycle_time;
148
149      }
```

```
150  }
151
152  /***************************************************/
153
154  /* Define the activities for the Slow_Thread */
155
156  void Slow_Thread_entry(ULONG thread_input)
157  {
158  UINT status;
159  ULONG current_time;
160  ULONG start_time, cycle_time;
161
162      while(1)
163      {
164
165          /* Get the starting time for this cycle */
166          start_time = tx_time_get();
167
168          /* Activity 5: 12 timer-ticks *** critical section ***
169          Get the mutex with suspension. */
170
171          status = tx_mutex_get(&my_mutex, TX_WAIT_FOREVER);
172          if (status != TX_SUCCESS) break; /* Check status */
173
174          tx_thread_sleep(12);
175
176          /* Release the mutex. */
177          status = tx_mutex_put(&my_mutex);
178          if (status != TX_SUCCESS) break; /* Check status */
179
180          /* Activity 6: 8 timer-ticks. */
181          tx_thread_sleep(8);
182
183          /* Activity 7: 11 timer-ticks *** critical section ***
184          Get the mutex with suspension. */
185
186          status = tx_mutex_get(&my_mutex, TX_WAIT_FOREVER);
187          if (status != TX_SUCCESS) break; /* Check status */
188
189          tx_thread_sleep(11);
```

```
190
191        /* Release the mutex. */
192        status = tx_mutex_put(&my_mutex);
193        if (status != TX_SUCCESS) break; /* Check status */
194
195        /* Activity 8: 9 timer-ticks. */
196        tx_thread_sleep(9);
197
198        /* Increment thread counter, compute cycle time & total
              time */
199        Slow_Thread_counter++;
200        current_time = tx_time_get();
201        cycle_time = current_time-start_time;
202        total_slow_time = total_slow_time + cycle_time;
203
204    }
205 }
206
207 /****************************************************/
208
209 /* Display statistics at periodic intervals */
210 void print_stats (ULONG invalue)
211 {
212    ULONG current_time, avg_slow_time, avg_speedy_time;
213
214    if ((Speedy_Thread_counter>0) && (Slow_Thread_counter>0))
215    {
216        current_time = tx_time_get();
217        avg_slow_time = total_slow_time/Slow_Thread_counter;
218        avg_speedy_time = total_speedy_time/Speedy_Thread_counter;
219
220        printf("\n**** Timing Info Summary\n\n");
221        printf("Current Time:            %lu\n", current_time);
222        printf(" Speedy_Thread counter: %lu\n",
                                        Speedy_Thread_counter);
223        printf(" Speedy_Thread avg time: %lu\n", avg_speedy_time);
224        printf(" Slow_Thread counter:   %lu\n",
                                        Slow_Thread_counter);
225        printf(" Slow_Thread avg time:  %lu\n\n",
                                        avg_slow_time);
```

```
226    }
227    else printf("Bypassing print_stats, Time: %lu\n",
       tx_time_get());
228  }
```

10.13 Application Timer Internals

When the TX_TIMER data type is used to declare an application timer, a Timer Control Block is created, and that Control Block is added to a doubly linked circular list, as illustrated in Figure 10.16.

The pointer named tx_timer_created_ptr points to the first Control Block in the list. See the fields in the Timer Control Block for timer attributes, values, and other pointers.

To quickly determine when the next timer will expire, ThreadX maintains an array of active timer linked-list head pointers, as illustrated in Figure 10.17. Each head pointer points to a linked list of Timer Control Blocks, such as illustrated in Figure 10.16.

The name of the array of active timer head pointers is _tx_timer_list. There is a pointer named _tx_timer_current_ptr that moves to the next position in this array at every timer-tick (in wrap-around fashion). Every time this pointer moves to a new array position, all the timers in the corresponding linked list are processed.[6] The actual expiration values are

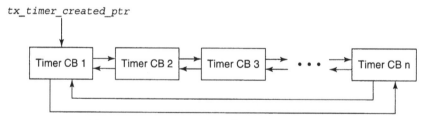

Figure 10.16: Created application timer list

---

[6]Note that "processed" does not necessarily mean that all the timers on the list have expired. Timers on the list that have an expiration value greater than the size of the list simply have their expiration value decremented by the list size and are then reinserted in the list. Thus, each timer list will be processed within a single timer-tick, but some timers may be only partially processed because they are not due to expire.

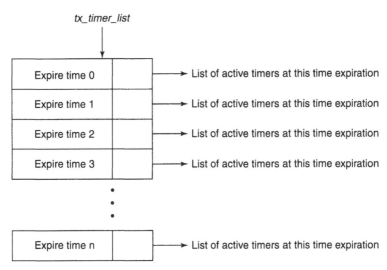

Figure 10.17: Example of array of active timer head pointers

"hashed" to the array index so there is no searching involved, thus optimizing processing speed. Thus, the next timer to expire will always be in the list pointed to by _tx_timer_ current_ptr. For more information about this process, see the files named tx_timer.h, tx_timer_interrupt.c, and tx_timer_interrupt.s.

10.14 Overview

The internal system clock is essential for capturing timing information, as illustrated by the sample program in this chapter.

ThreadX provides two services for interacting with the internal system clock—getting the current time and setting the clock to a new value. There is only one internal system clock; by contrast, applications may create an unlimited number of application timers.

A one-shot application timer expires once, executes its expiration function, and then terminates. A periodic timer expires repeatedly until it is stopped, with each expiration resulting in a call to its expiration function. A periodic timer has an initial expiration value, and a reschedule value for continuing expirations.

When a timer is created, it can be activated immediately, or it can be activated at some later time by a thread.

Eight services are available for use with application timers: *create, activate, change, deactivate, retrieve information* (3 services), *and delete*.

10.15 Key Terms and Phrases

activation of timer	internal system clock
application timer	one-shot timer
Application Timer Control Block	periodic interrupt
application timer services	periodic timer
compute timing performance	system clock services
creating a timer	tick counter
deactivation of timer	timeout function
deleting a timer	timer activation options
expiration function	timer interrupt
expiration time	timer-tick

10.16 Problems

1. Describe a scenario in which you would use the internal system clock rather than an application timer.

2. Describe a scenario in which you would use an application timer rather than the internal system clock.

3. If an application timer is created with the TX_AUTO_ACTIVATE option, when will that timer be activated? If that timer is created with the TX_NO_ACTIVATE option, when will that timer become activated?

4. When an application timer is created, the initial number of timer-ticks must be greater than zero. Given a timer that has already been created and deactivated, how would you cause its expiration function print_stats to execute at some arbitrary time in the future?

5. Assume that application timer my_timer and variable status have been declared. What will happen as a result of the following timer creation service call?

```
status = tx_timer_create(&my_timer,"my_timer_name",
                         my_timer_function, 0X1234,
                         100, 0, TX_NO_ACTIVATE);
```

6. Assume that application timer my_timer has been created as in the previous problem. Inspect the following fields in its Control Block: tx_remaining_ticks and tx_re_initialize_ticks.

7. The section titled "Sample System Using Timers to Measure Thread Performance" presents most of the needed modifications from the previous chapter's examples. Complete and execute this sample system.

8. Assume that application timer my_timer has been created. What services would you use to change that timer to a one-shot timer that expires exactly 70 timer-ticks from some arbitrary current time?

9. Assume that application timer my_timer has been created. What services would you use to carry out the following operations: if the remaining number of timer-ticks exceeds 60, then change the timer so that it expires in exactly 2 timer-ticks, and every 75 timer-ticks after the first expiration.

10. Suppose you want to create one application timer that expires exactly at the following times: 50, 125, 150, and 210. How would you do this? State your assumptions.

11. In general, which method is better for obtaining new timer behavior: (1) delete an application timer, then create a new timer with the same name and with new characteristics, or (2) deactivate that timer and change its characteristics?

CHAPTER 11

Event Notification and Synchronization with Counting Semaphores

11.1 Introduction

ThreadX provides 32-bit counting semaphores with counts that range in value from 0 to $2^{32} - 1$, or 4,294,967,295 (inclusive). There are two operations that affect the values of counting semaphores: tx_semaphore_get and tx_semaphore_put. The get operation decreases the semaphore by one. If the semaphore is 0, the get operation fails. The inverse of the get operation is the put operation, which increases the semaphore by one.

Each counting semaphore is a public resource. ThreadX imposes no constraints as to how counting semaphores are used.

An *instance* of a counting semaphore is a single count. For example, if the count is five, then that semaphore has five instances. Similarly, if the count is zero, then that semaphore has no instances. The get operation takes one instance from the counting semaphore by decrementing its count. Similarly, the put operation places one instance in the counting semaphore by incrementing its count.

Like mutexes, counting semaphores are often used for mutual exclusion. One major difference between these objects is that counting semaphores do not support ownership, a concept that is central to mutexes. Even so, counting semaphores are more versatile; they can also be used for event notification and interthread synchronization.

Mutual exclusion pertains to controlling threads' access to certain application areas (typically, *critical sections* and *application resources*).[1] When used for mutual exclusion, the "current count" of a semaphore represents the total number of threads that are allowed access to that semaphore's associated resource. In most cases, counting semaphores used for mutual exclusion have an initial value of 1, meaning that only one thread can access the associated resource at a time. Counting semaphores that have values restricted to 0 or 1 are commonly called *binary semaphores*.

If a binary semaphore is used, the user must prevent that same thread from performing a get operation on a semaphore it already controls. A second get would fail and could suspend the calling thread indefinitely, as well as make the resource permanently unavailable.

Counting semaphores can also be used for event notification, as in a *producer-consumer* application. In this application, the consumer attempts to get the counting semaphore before "consuming" a resource (such as data in a queue); the producer increases the semaphore count whenever it makes something available. In other words, the producer *places* instances in the semaphore and the consumer attempts to *take* instances from the semaphore. Such semaphores usually have an initial value of 0 and do not increase until the producer has something ready for the consumer.

Applications can create counting semaphores either during initialization or during run-time. The initial count of the semaphore is specified during creation. An application may use an unlimited number of counting semaphores.

Application threads can suspend while attempting to perform a get operation on a semaphore with a current count of zero (depending on the value of the wait option).

When a put operation is performed on a semaphore and a thread suspended on that semaphore, the suspended thread completes its get operation and resumes. If multiple threads are suspended on the same counting semaphore, they resume in the same order they occur on the suspended list (usually in FIFO order).

An application can cause a higher-priority thread to be resumed first, if the application calls tx_semaphore_prioritize prior to a semaphore put call. The semaphore prioritization

---

[1]Review the discussion about mutual exclusion in Chapter 8 dealing with mutexes.

service places the highest-priority thread at the front of the suspension list, while leaving all other suspended threads in the same FIFO order.

11.2 Counting Semaphore Control Block

The characteristics of each counting semaphore are found in its Control Block.[2] It contains useful information such as the current semaphore count and the number of threads suspended for this counting semaphore. Counting Semaphore Control Blocks can be located anywhere in memory, but it is most common to make the Control Block a global structure by defining it outside the scope of any function. Figure 11.1 contains many of the fields that comprise this Control Block.

A Counting Semaphore Control Block is created when a counting semaphore is declared with the TX_SEMAPHORE data type. For example, we declare my_semaphore as follows:

```
TX_SEMAPHORE my_semaphore;
```

The declaration of counting semaphores normally appears in the declaration and definition section of the application program.

Field	Description
tx_semaphore_id	Counting Semaphore ID
tx_semaphore_name	Pointer to counting semaphore name
tx_semaphore_count	Actual semaphore count
tx_semaphore_suspension_list	Pointer to counting semaphore suspension list
tx_semaphore_suspended_count	Number of threads suspended for this semaphore
tx_semaphore_created_next	Pointer to the next semaphore in the created list
tx_semaphore_created_previous	Pointer to the previous semaphore in the created list

Figure 11.1: Counting Semaphore Control Block

---

[2]The structure of the Counting Semaphore Control Block is defined in the **tx_api.h** file.

11.3 Avoiding Deadly Embrace

One of the most dangerous pitfalls associated in using semaphores for mutual exclusion is the so-called *deadly embrace*. A deadly embrace, or *deadlock,* is a condition in which two or more threads are suspended indefinitely while attempting to get semaphores already owned by the other threads. Refer to the discussion in Chapter 8 to find remedies for deadly embrace. This discussion applies to the counting semaphore object as well.

11.4 Avoiding Priority Inversion

Another pitfall associated with mutual exclusion semaphores is *priority inversion,* which was also discussed in Chapters 7 and 8. The groundwork for trouble is laid when a lower-priority thread acquires a mutual exclusion semaphore that a higher-priority thread needs. This sort of priority inversion in itself is normal. However, if threads that have intermediate priorities acquire the semaphore, the priority inversion may last for a nondeterministic amount of time. You can prevent this by carefully selecting thread priorities, by using preemption-threshold, and by temporarily raising the priority of the thread that owns the resource to that of the high-priority thread. Unlike mutexes, however, counting semaphores do not have a priority inheritance feature.

11.5 Summary of Counting Semaphore Services

Appendix G contains detailed information about counting semaphore services. This appendix contains information about each service such as the prototype, a brief description of the service, required parameters, return values, notes and warnings, allowable invocation, and an example showing how the service can be used. Figure 11.2 contains a listing of all available counting semaphore services. We will investigate each of these services in subsequent sections of this chapter.

11.6 Creating a Counting Semaphore

A counting semaphore is declared with the TX_SEMAPHORE data type and is defined with the tx_semaphore_create service. When defining a counting semaphore, you must specify its Control Block, the name of the counting semaphore, and the initial count for the semaphore. Figure 11.3 lists the attributes of a counting semaphore. The value for the count must be in the range from 0x00000000 to 0xFFFFFFFF (inclusive).

Counting semaphore service	Description
tx_semaphore_ceiling_put	Place an instance in counting semaphore with ceiling
tx_semaphore_create	Create a counting semaphore
tx_semaphore_delete	Delete a counting semaphore
tx_semaphore_get	Get an instance from a counting semaphore
tx_semaphore_info_get	Retrieve information about a counting semaphore
tx_semaphore_performance_info_get	Get semaphore performance information
tx_semaphore_performance_system_info_get	Get semaphore system performance information
tx_semaphore_prioritize	Prioritize the counting semaphore suspension list
tx_semaphore_put	Place an instance in a counting semaphore
tx_semaphore_put_notify	Notify application when semaphore is put

Figure 11.2: Services of the counting semaphore

Counting semaphore control block
Counting semaphore name
Initial count

Figure 11.3: Counting semaphore attributes

Figure 11.4 illustrates the use of this service to create a counting semaphore. We give our counting semaphore the name "my_semaphore" and we give it an initial value of one. As noted before, this is typically the manner in which a *binary semaphore* is created.

If the variable *status* contains the return value of TX_SUCCESS, we have successfully created a counting semaphore.

11.7 Deleting a Counting Semaphore

Use the tx_semaphore_delete service to delete a counting semaphore. All threads that have been suspended because they are waiting for a semaphore instance are resumed and

```
TX_SEMAPHORE my_semaphore;
UINT status;

/* Create a counting semaphore with an initial value of 1.
   This is typically the technique used to create a binary
   semaphore. Binary semaphores are used to provide
   protection over a common resource. */
status = tx_semaphore_create(&my_semaphore,
                           "my_semaphore_name", 1);

/* If status equals TX_SUCCESS, my_semaphore is
   ready for use. */
```

Figure 11.4: Creating a counting semaphore

```
TX_SEMAPHORE my_semaphore;
UINT status;

...

/* Delete counting semaphore. Assume that the
   counting semaphore has already been created. */

status = tx_semaphore_delete(&my_semaphore);

/* If status equals TX_SUCCESS, the counting
   semaphore has been deleted. */
```

Figure 11.5: Deleting a counting semaphore

receive a TX_DELETED return status. Make certain that you don't try to use a deleted semaphore. Figure 11.5 shows how a counting semaphore can be deleted.

If variable *status* contains the return value TX_SUCCESS, we have successfully deleted the counting semaphore.

11.8 Getting an Instance of a Counting Semaphore

The tx_semaphore_get service retrieves an instance (a single count) from the specified counting semaphore. If this call succeeds, the semaphore count decreases by one. Figure 11.6 shows how to get an instance of a counting semaphore, where we use the wait option value TX_WAIT_FOREVER.

If variable *status* contains the return value TX_SUCCESS, we have successfully obtained an instance of the counting semaphore called my_semaphore.

```
TX_SEMAPHORE my_semaphore;
UINT status;

...

/* Get a semaphore instance from the semaphore
   "my_semaphore." If the semaphore count is zero,
   suspend until an instance becomes available.
   Note that this suspension is only possible from
   application threads. */
status = tx_semaphore_get(&my_semaphore,
                    TX_WAIT_FOREVER);

/* If status equals TX_SUCCESS, the thread has
   obtained an instance of the semaphore. */
```

Figure 11.6: Get an instance from a counting semaphore

11.9 Retrieving Information about Counting Semaphores

There are three services that enable you to retrieve vital information about semaphores. The first such service for semaphores—the tx_semaphore_info_get service—retrieves a subset of information from the Semaphore Control Block. This information provides a "snapshot" at a particular instant in time, i.e., when the service is invoked. The other two services provide summary information that is based on the gathering of run-time performance data. One service—the tx_semaphore_performance_info_get service—provides an information summary for a particular semaphore up to the time the service is invoked. By contrast the tx_semaphore_performance_system_info_get retrieves an information summary for all semaphores in the system up to the time the service is invoked. These services are useful in analyzing the behavior of the system and determining whether there are potential problem areas. The tx_semaphore_info_get[3] service retrieves several useful pieces of information about a counting semaphore. The information that is retrieved includes the counting semaphore name, its current count, the number of threads suspended for this semaphore, and a pointer to the next created counting semaphore. Figure 11.7 shows how this service can be used to obtain information about a counting semaphore.

---

[3]By default, only the *tx_semaphore_info_get* service is enabled. The other two information gathering services must be enabled in order to use them.

```
TX_SEMAPHORE my_semaphore;
CHAR *name;
ULONG current_value;
TX_THREAD *first_suspended;
ULONG suspended_count;
TX_SEMAPHORE *next_semaphore;
UINT status;
…

/* Retrieve information about the previously
   created semaphore "my_semaphore." */

status = tx_semaphore_info_get(&my_semaphore, &name,
                               &current_value,
                               &first_suspended,
                               &suspended_count,
                               &next_semaphore);
/* If status equals TX_SUCCESS, the information
   requested is valid. */
```

Figure 11.7: Get information about a counting semaphore

If variable *status* contains the return value TX_SUCCESS, we have obtained valid information about the counting semaphore called my_semaphore.

11.10 Prioritizing a Counting Semaphore Suspension List

When a thread is suspended because it is waiting for a counting semaphore, it is placed in the suspension list in a FIFO manner. When a counting semaphore instance becomes available, the first thread in that suspension list (regardless of priority) obtains ownership of that instance. The tx_semaphore_prioritize service places the highest-priority thread suspended on a specific counting semaphore at the front of the suspension list. All other threads remain in the same FIFO order in which they were suspended. Figure 11.8 shows how this service can be used.

If variable *status* contains the return value TX_SUCCESS, the highest-priority thread suspended on an instance of my_semaphore has been placed at the front of the suspension list.

11.11 Placing an Instance in a Counting Semaphore

The tx_semaphore_put service places an instance in a counting semaphore, i.e., it increases the count by one. If there is a thread suspended on this semaphore when the put

```
TX_SEMAPHORE my_semaphore;
UINT status;

/* Ensure that the highest priority thread will
   receive the next instance of this semaphore. */

...

status = tx_semaphore_prioritize(&my_semaphore);

/* If status equals TX_SUCCESS, the highest priority
   suspended thread is at the front of the list. The
   next tx_semaphore_put call made to this semaphore
   will wake up this thread. */
```

Figure 11.8: Prioritize the counting semaphore suspension list

```
TX_SEMAPHORE my_semaphore;
UINT status;

...

/* Increment the counting semaphore "my_semaphore." */

status = tx_semaphore_put(&my_semaphore);

/* If status equals TX_SUCCESS, the semaphore count has
   been incremented. Of course, if a thread was waiting,
   it was given the semaphore instance and resumed. */
```

Figure 11.9: Place an instance on a counting semaphore

service is performed, the suspended thread's get operation completes and that thread is resumed. Figure 11.9 shows how this service can be used.

If variable *status* contains the return value TX_SUCCESS, the semaphore count has been incremented (an instance has been placed in the semaphore). If a thread was suspended for this semaphore, then that thread receives this instance and resumes execution.

11.12 Placing an Instance in a Semaphore Using a Ceiling

The tx_semaphore_put service places an instance in a counting semaphore without regard to the current count of the semaphore. However, if you want to make certain that

```
TX_SEMAPHORE my_semaphore;
UINT status;

...

/* Increment the semaphore "my_semaphore" using a ceiling */

status = tx_semaphore_ceiling_put(&my_semaphore, 3);

/* If status equals TX_SUCCESS, the semaphore count has
   been incremented. Of course, if a thread was waiting,
   it was given the semaphore instance and resumed.

   If status equals TX_CEILING_EXCEEDED than the semaphore is not
   incremented because the resulting value would have been greater
   Than the ceiling value of 3. */
```

Figure 11.10: Place an instance on a semaphore using a ceiling

the count is always less than or equal to a certain value, use the tx_semaphore_ceiling_ put service. When this service is invoked, the semaphore is incremented only if the resulting count would be less than or equal to the ceiling value. Figure 11.10 shows how this service can be used.

11.13 Semaphore Notification and Event-Chaining[4]

The tx_semaphore_put_notify service registers a notification callback function that is invoked whenever an instance is placed on the specified semaphore. The processing of the notification callback is defined by the application. This is an example of event-chaining where notification services are used to chain various synchronization events together. This is typically useful when a single thread must process multiple synchronization events.

11.14 Comparing a Counting Semaphore with a Mutex

A counting semaphore resembles a mutex in several respects, but there are differences, as well as reasons to use one resource over the other. Figure 11.11 reproduces the comparison chart for these two objects, which first appeared in Chapter 4.

---

[4]Event-chaining is a trademark of Express Logic, Inc.

	Mutex	Counting Semaphore
Speed	Somewhat slower than a semaphore	Semaphore is generally faster than a mutex and requires fewer system resources
Thread ownership	Only one thread can own a mutex	No concept of thread ownership for a semaphore – any thread can decrement a counting semaphore if its current count exceeds zero
Priority inheritance	Available only with a mutex	Feature not available for semaphores
Mutual exclusion	Primary purpose of a mutex – a mutex should be used only for mutual exclusion	Can be accomplished with the use of a binary semaphore, but there may be pitfalls
Inter-thread synchronization	Do not use a mutex for this purpose	Can be performed with a semaphore, but an event flags group should be considered also
Event notification	Do not use a mutex for this purpose	Can be performed with a semaphore
Thread suspension	Thread can suspend if another thread already owns the mutex (depends on value of wait option)	Thread can suspend if the value of a counting semaphore is zero (depends on value of wait option)

Figure 11.11: Comparison of a mutex with a counting semaphore

A mutex is exceptionally robust in providing mutual exclusion. If this is crucial to your application, then using a mutex is a good decision. However, if mutual exclusion is not a major factor in your application, then use a counting semaphore because it is slightly faster and uses fewer system resources.

To illustrate the use of a counting semaphore, we will replace a mutex with a binary semaphore in the next sample system.

11.15 Sample System Using a Binary Semaphore in Place of a Mutex

This sample system is a modification of the one discussed in the preceding chapter. The only goal here is to replace the mutex from that system with a binary semaphore. We will retain the timing facilities of that system to compare the results of thread processing by using a mutex versus using a binary semaphore.

Figure 11.12 shows a modification of the Speedy_Thread activities, in which we have replaced references to a mutex with references to a binary semaphore. The priorities and times remain the same as in the previous system. The shaded boxes represent the critical sections.

Figure 11.13 shows a modification of the Slow_Thread activities. The only change we have made is to replace references to mutexes with references to binary semaphores.

In Chapter 10, we created a sample system that produced output that began as follows:

```
**** Timing Info Summary
Current Time: 500
Speedy_Thread counter: 22
Speedy_Thread avg time: 22
Slow_Thread counter: 11
Slow_Thread avg time: 42
```

We want our new sample system to perform the same operations as the previous system. We will discuss a series of changes to be applied to the previous system so that all references to a mutex will be replaced with references to a binary semaphore. The

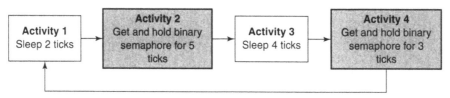

Figure 11.12: Activities of the Speedy_Thread (priority = 5)

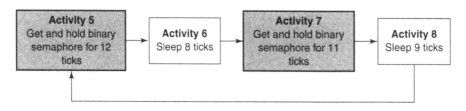

Figure 11.13: Activities of the Slow_Thread (priority = 15)

complete program listing, called *11a_sample_system.c,* is located in the next section of this chapter and on the attached CD.

The first change occurs in the declaration and definitions section of our program, where we replace the declaration of a mutex with the declaration of a binary semaphore, as follows.

```
TX_SEMAPHORE my_semaphore;
```

A binary semaphore is a special case of a counting semaphore, so the declaration of each is the same. The next change occurs in the application definitions section of our program, where we replace the creation of a mutex with the creation of a binary semaphore, as follows:

```
/* Create the binary semaphore used by both threads */
tx_semaphore_create(&my_semaphore, "my_semaphore", 1);
```

There are two primary differences between the definition of a mutex and the definition of a binary semaphore. First, only mutexes support priority inheritance, so that option does not appear in the argument list for semaphore creation. Second, only semaphores have counts, so the argument list must include an initial value. In the above semaphore creation, the initial count is one (1), which is the most commonly used initial value for a binary semaphore.[5] This means that the binary semaphore has one instance available that may be obtained by a thread.

The remaining changes occur in the function definitions section of our program. We need to change all references to mutexes to binary semaphores to protect critical sections in Activities 2 and 4 for the Speedy_Thread, and Activities 5 and 7 for the Slow_Thread. We will show only the changes for the Speedy_Thread and will leave the Slow_Thread changes as an exercise for the reader. Figure 11.14 contains the necessary changes for Activity 2. Most of the modifications involve changing references to a mutex with references to a binary semaphore.

Figure 11.15 contains the necessary changes for Activity 4. Most of the modifications involve changing references to a mutex with references to a binary semaphore.

---

[5]The only other possible value is zero (0). It is rarely used as an initial value for a binary semaphore.

```
/* Activity 2:  5 timer-ticks   *** critical section ***
Get an instance of the binary semaphore with suspension. */

status = tx_semaphore_get(&my_semaphore, TX_WAIT_FOREVER);
if (status != TX_SUCCESS)  break;  /* Check status */

tx_thread_sleep(5);

/* Place an instance in the binary semaphore. */
status = tx_semaphore_put(&my_semaphore);
if (status != TX_SUCCESS)  break;  /* Check status */
```

Figure 11.14: Changes to Activity 2

```
/* Activity 4:  3 timer-ticks   *** critical section ***
Get an instance of the binary semaphore with suspension. */

status = tx_semaphore_get(&my_semaphore, TX_WAIT_FOREVER);
if (status != TX_SUCCESS)  break;  /* Check status */

tx_thread_sleep(3);

/* Place an instance in the binary semaphore. */
status = tx_semaphore_put(&my_semaphore);
if (status != TX_SUCCESS)  break;  /* Check status */
```

Figure 11.15: Changes to Activity 4

11.16 Listing for 11a_sample_system.c

The sample system named *11a_sample_system.c* is located on the attached CD. The complete listing appears below; line numbers have been added for easy reference.

```
001  /* 11a_sample_system.c
002
003     Create two threads, and one mutex.
004     Use arrays for the thread stacks.
005     A binary semaphore protects the critical sections.
006     Use an application timer to display thread timings. */
007
008  /*************************************************/
009  /*     Declarations, Definitions, and Prototypes     */
010  /*************************************************/
011
```

```
012
013  #include "tx_api.h"
014  #include <stdio.h >
015
016  #define    STACK_SIZE     1024
017
018  CHAR stack_speedy[STACK_SIZE];
019  CHAR stack_slow[STACK_SIZE];
020
021
022  /* Define the ThreadX object control blocks... */
023
024  TX_THREAD Speedy_Thread;
025  TX_THREAD Slow_Thread;
026
027  TX_SEMAPHORE my_semaphore;
028
029  /* Declare the application timer */
030  TX_TIMER stats_timer;
031
032  /* Declare the counters and accumulators */
033  ULONG Speedy_Thread_counter = 0,
034       total_speedy_time = 0;
035  ULONG Slow_Thread_counter = 0,
036       total_slow_time = 0;
037
038  /* Define prototype for expiration function */
039  void print_stats(ULONG);
040
041  /* Define thread prototypes. */
042
043  void Speedy_Thread_entry(ULONG thread_input);
044  void Slow_Thread_entry(ULONG thread_input);
045
046  /******************************************************/
047  /*               Main Entry Point                   */
048  /******************************************************/
049
050  /* Define main entry point. */
051
```

```
052  int main()
053  {
054
055      /* Enter the ThreadX kernel. */
056      tx_kernel_enter();
057  }
058
059  /****************************************************/
060  /*              Application Definitions            */
061  /****************************************************/
062
063  /* Define what the initial system looks like. */
064
065  void tx_application_define(void *first_unused_memory)
066  {
067
068
069      /* Put system definitions here,
070      e.g., thread and semaphore creates */
071
072      /* Create the Speedy_Thread. */
073      tx_thread_create(&Speedy_Thread, "Speedy_Thread",
074                       Speedy_Thread_entry, 0,
075                       stack_speedy, STACK_SIZE,
076                       5, 5, TX_NO_TIME_SLICE, TX_AUTO_START);
077
078      /* Create the Slow_Thread */
079      tx_thread_create(&Slow_Thread, "Slow_Thread",
080                       Slow_Thread_entry, 1,
081                       stack_slow, STACK_SIZE,
082                       15, 15, TX_NO_TIME_SLICE, TX_AUTO_START);
083
084      /* Create the binary semaphore used by both threads */
085      tx_semaphore_create(&my_semaphore, "my_semaphore", 1);
086
087      /* Create and activate the timer */
088      tx_timer_create (&stats_timer, "stats_timer", print_stats,
089                       0x1234, 500, 500, TX_AUTO_ACTIVATE);
090
091  }
```

```
092
093    /****************************************************/
094    /*                Function Definitions              */
095    /****************************************************/
096
097    /* Define the activities for the Speedy_Thread */
098
099    void Speedy_Thread_entry(ULONG thread_input)
100    {
101    UINT status;
102    ULONG current_time;
103    ULONG start_time, cycle_time;
104
105       while(1)
106       {
107
108          /* Get the starting time for this cycle */
109          start_time = tx_time_get();
110
111          /* Activity 1: 2 timer-ticks. */
112          tx_thread_sleep(2);
113
114          /* Activity 2: 5 timer-ticks *** critical section ***
115          Get an instance of the binary semaphore with
             suspension. */
116
117          status = tx_semaphore_get(&my_semaphore,
             TX_WAIT_FOREVER);
118          if (status ! = TX_SUCCESS) break; /* Check status */
119
120          tx_thread_sleep(5);
121
122          /* Place an instance in the binary semaphore. */
123          status = tx_semaphore_put(&my_semaphore);
124          if (status ! = TX_SUCCESS) break; /* Check status */
125
126          /* Activity 3: 4 timer-ticks. */
127          tx_thread_sleep(4);
128
129          /* Activity 4: 3 timer-ticks *** critical section ***
```

```
130        Get an instance of the binary semaphore with
              suspension. */
131
132        status = tx_semaphore_get(&my_semaphore,
              TX_WAIT_FOREVER);
133        if (status ! = TX_SUCCESS) break; /* Check status */
134
135        tx_thread_sleep(3);
136
137        /* Place an instance in the binary semaphore. */
138        status = tx_semaphore_put(&my_semaphore);
139        if (status ! = TX_SUCCESS) break; /* Check status */
140
141        /* Increment thread counter, compute cycle time & total
              time */
142        Speedy_Thread_counter++;
143        current_time = tx_time_get();
144        cycle_time = current_time - start_time;
145        total_speedy_time = total_speedy_time + cycle_time;
146
147    }
148 }
149
150 /****************************************************/
151
152 /* Define the activities for the Slow_Thread */
153
154 void Slow_Thread_entry(ULONG thread_input)
155 {
156 UINT status;
157 ULONG current_time;
158 ULONG start_time, cycle_time;
159
160 while(1)
161 {
162
163    /* Get the starting time for this cycle */
164    start_time = tx_time_get();
165
166    /* Activity 5: 12 timer-ticks *** critical section ***
167     Get an instance of the binary semaphore with suspension. */
```

```
168
169    status = tx_semaphore_get(&my_semaphore, TX_WAIT_FOREVER);
170    if (status ! = TX_SUCCESS) break; /* Check status */
171
172    tx_thread_sleep(12);
173
174    /* Place an instance in the binary semaphore. */
175    status = tx_semaphore_put(&my_semaphore);
176    if (status ! = TX_SUCCESS) break; /* Check status */
177
178    /* Activity 6: 8 timer-ticks. */
179    tx_thread_sleep(8);
180
181    /* Activity 7: 11 timer-ticks *** critical section ***
182    Get an instance of the binary semaphore with suspension. */
183
184    status = tx_semaphore_get(&my_semaphore, TX_WAIT_FOREVER);
185    if (status ! = TX_SUCCESS) break; /* Check status */
186
187    tx_thread_sleep(11);
188
189    /* Place an instance in the binary semaphore. */
190    status = tx_semaphore_put(&my_semaphore);
191    if (status ! = TX_SUCCESS) break; /* Check status */
192
193    /* Activity 8: 9 timer-ticks. */
194    tx_thread_sleep(9);
195
196    /* Increment thread counter, compute cycle time & total
          time */
197    Slow_Thread_counter + +;
198    current_time = tx_time_get();
199    cycle_time = current_time - start_time;
200    total_slow_time = total_slow_time + cycle_time;
201
202    }
203 }
204
205 /****************************************************/
206
207 /* Display statistics at periodic intervals */
```

```
208
209   void print_stats (ULONG invalue)
210   {
211     ULONG current_time, avg_slow_time, avg_speedy_time;
212
213     if ((Speedy_Thread_counter>0) && (Slow_Thread_counter>0))
214     {
215        current_time = tx_time_get();
216        avg_slow_time = total_slow_time / Slow_Thread_counter;
217        avg_speedy_time = total_speedy_time /
                Speedy_Thread_counter;
218
219        printf("\n**** Timing Info Summary\n\n");
220        printf("Current Time:              %lu\n", current_time);
221        printf(" Speedy_Thread counter:  %lu\n",
                Speedy_Thread_counter);
222        printf(" Speedy_Thread avg time: %lu\n", avg_speedy_time);
223        printf(" Slow_Thread counter:    %lu\n",
                Slow_Thread_counter);
224        printf(" Slow_Thread avg time:   %lu\n\n", avg_slow_time);
225     }
226     else printf("Bypassing print_stats, Time: %lu\n",
            tx_time_get());
227   }
```

11.17 Sample System Using a Counting Semaphore in a Producer-Consumer Application

Counting semaphores are used primarily for mutual exclusion, event notification, or synchronization. We used a counting semaphore for mutual exclusion in the previous sample system; we will use a counting semaphore for event notification in this system. We will modify the previous system to achieve this purpose by creating a *producer-consumer* application. The Speedy_Thread will act as the producer and the Slow_Thread will act as the consumer. The Speedy_Thread will place instances in the counting semaphore (i.e., increment the semaphore count) and the Slow_Thread will wait for an instance in the semaphore and then take it (i.e., decrement the semaphore count). The counting semaphore simulates a storage facility, as illustrated by Figure 11.16. In this case, the facility just stores instances of the semaphore. In other applications, it could

Figure 11.16: Producer-consumer system

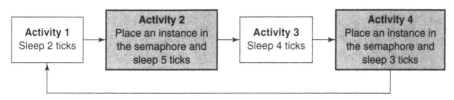

Figure 11.17: Activities of the producer (Speedy_Thread) where priority = 5

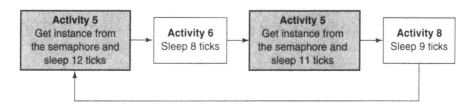

Figure 11.18: Activities of the consumer (Slow_Thread) where priority = 15

store data bytes, Internet packets, or practically anything. The application logic to use the semaphores remains the same regardless of what is stored.

Figure 11.17 contains a modification of the activities for the Speedy_Thread, which serves as the producer in this system. The producer thread contains no critical sections. We will use the same activity times as the previous system.

Figure 11.18 contains a modification of the activities for the Slow_Thread, which serves as the consumer in this system. The consumer thread contains no critical sections either. We will use the same activity times as the previous system.

If the consumer attempts to get an instance from the counting semaphore and no instances are available, then the consumer waits (suspends) until an instance becomes available.

This system provides a good example of event notification. The producer communicates with the consumer via the counting semaphore. The producer creates an event notification by placing an instance in the counting semaphore. The consumer, in effect, waits until this event notification is issued before getting an instance from the counting semaphore.

Following is a portion of the sample output that we would expect to be displayed from the producer-consumer system.

Producer-Consumer System–Timing Summary	
Current Time:	500
Speedy_Thread counter:	35
Speedy_Thread avg time:	14
Slow_Thread counter:	12
Slow_Thread avg time:	40
Producer-Consumer System–Timing Summary	
Current Time:	1000
Speedy_Thread counter:	71
Speedy_Thread avg time:	14
Slow_Thread counter:	24
Slow_Thread avg time:	40

We will use the same application timer and expiration function as the previous system. We will modify that system so that the Speedy_Thread becomes the producer and the Slow_Thread becomes the consumer, according to Figure 11.17 and Figure 11.18.

The first change occurs in the application definition section of our program, where the binary semaphore is changed to a counting semaphore. Figure 11.19 shows this change.

Changing the initial value of the semaphore from 1 to 0 is the only change that appears in the previous figure. This emphasizes the difference between a binary semaphore, which is restricted to the values 1 and 0, and a counting semaphore, which has a count range of 0 to 0xFFFFFFFF, inclusive.

```
/* Create the counting semaphore used by both threads  */
tx_semaphore_create(&my_semaphore, "my_semaphore", 0);
```

Figure 11.19: Creating the counting semaphore for the producer-consumer system

There are no critical sections in the producer-consumer system, so we must remove that protection wherever it occurs, i.e., in Activities 2, 4, 5, and 7. A critical section could be protected with a *get semaphore/put semaphore* pair. However, the producer will use only *put semaphore* operations for Activities 2 and 4. Conversely, the consumer will use only *get semaphore* operations for Activities 5 and 7—neither thread attempts to get and then put the semaphore. Figure 11.20 contains the necessary changes to Activity 2 of the producer.

The producer will always be able to place an instance on the counting semaphore. Activity 4 is similar to Activity 2, so we leave its changes as an exercise for the reader. Figure 11.21 contains the necessary changes to Activity 5 of the consumer.

The consumer must wait for an instance from the counting semaphore if one is not available. Activity 7 is similar to Activity 5, so we leave its changes as an exercise for the reader as well. The next section contains a complete listing of this system.

11.18 Listing for 11b_sample_system.c

The sample system named *11b_sample_system.c* is located on the attached CD. The complete listing appears below; line numbers have been added for easy reference.

```
/* Activity 2:  5 timer-ticks.  */
/* Put an instance in the counting semaphore.  */

status = tx_semaphore_put (&my_semaphore);

if (status != TX_SUCCESS)  break;  /* Check status */
tx_thread_sleep(5);
```

Figure 11.20: Activity 2 of the producer

```
/* Activity 5 - get an instance of the counting
   semaphore with suspension and sleep 12 timer-ticks.  */

status = tx_semaphore_get (&my_semaphore, TX_WAIT_FOREVER);

if (status != TX_SUCCESS)  break;  /* Check status */
tx_thread_sleep(12);
```

Figure 11.21: Activity 5 of the consumer

```
001  /* 11b_sample_system.c
002
003     Producer-Consumer System
004
005     Create two threads and one counting semaphore.
006     Threads cooperate with each other via the semaphore.
007     Timer generates statistics at periodic intervals.
008     Producer (Speedy_Thread) - Consumer (Slow_Thread) */
009
010
011  /***************************************************/
012  /*     Declarations, Definitions, and Prototypes      */
013  /***************************************************/
014
015  #include "tx_api.h"
016  #include <stdio.h >
017
018  #define    STACK_SIZE    1024
019
020  /* Declare stacks for both threads. */
021  CHAR stack_speedy[STACK_SIZE];
022  CHAR stack_slow[STACK_SIZE];
023
024  /* Define the ThreadX object control blocks. */
025  TX_THREAD    Speedy_Thread;
026  TX_THREAD    Slow_Thread;
027  TX_SEMAPHORE my_semaphore;
028
029  /* Declare the application timer */
030  TX_TIMER stats_timer;
031
032  /* Declare the counters and accumulators */
033  ULONG Speedy_Thread_counter = 0,
034        total_speedy_time = 0;
035  ULONG Slow_Thread_counter = 0,
036        total_slow_time = 0;
037
038  /* Define thread prototypes. */
039  void Speedy_Thread_entry(ULONG thread_input);
040  void Slow_Thread_entry(ULONG thread_input);
```

```
041
042  /* Define prototype for expiration function */
043  void print_stats(ULONG);
044
045
046  /***************************************************/
047  /*               Main Entry Point                 */
048  /***************************************************/
049
050  /* Define main entry point. */
051
052  int main()
053  {
054
055      /* Enter the ThreadX kernel. */
056      tx_kernel_enter();
057  }
058
059
060  /***************************************************/
061  /*            Application Definitions              */
062  /***************************************************/
063
064  /* Define what the initial system looks like. */
065
066  void tx_application_define(void *first_unused_memory)
067  {
068
069      /* Put system definitions here,
070         e.g., thread, semaphore, and timer creates */
071
072      /* Create the Speedy_Thread. */
073      tx_thread_create(&Speedy_Thread, "Speedy_Thread",
074                       Speedy_Thread_entry, 0,
075                       stack_speedy, STACK_SIZE,
076                       5, 5, TX_NO_TIME_SLICE, TX_AUTO_START);
077
078      /* Create the Slow_Thread */
079      tx_thread_create(&Slow_Thread, "Slow_Thread",
080                       Slow_Thread_entry, 1,
```

```
081                        stack_slow, STACK_SIZE,
082                        15, 15, TX_NO_TIME_SLICE, TX_AUTO_START);
083
084     /* Create the counting semaphore used by both threads */
085     tx_semaphore_create(&my_semaphore, "my_semaphore", 0);
086
087     /* Create and activate the timer */
088     tx_timer_create (&stats_timer, "stats_timer", print_stats,
089                        0x1234, 500, 500, TX_AUTO_ACTIVATE);
090 }
091
092
093 /**************************************************/
094 /*               Function Definitions             */
095 /**************************************************/
096
097 /* Define the activities for the Producer (speedy) thread */
098
099 void Speedy_Thread_entry(ULONG thread_input)
100 {
101
102 UINT  status;
103 ULONG start_time, cycle_time, current_time;
104
105     while(1)
106     {
107      /* Get the starting time for this cycle */
108      start_time = tx_time_get();
109
110      /* Activity 1: 2 timer-ticks. */
111      tx_thread_sleep(2);
112
113      /* Put an instance in the counting semaphore. */
114      status = tx_semaphore_put (&my_semaphore);
115      if (status ! = TX_SUCCESS) break; /* Check status */
116
117      /* Activity 2: 5 timer-ticks. */
118      tx_thread_sleep(5);
119
```

```
120        /* Activity 3: 4 timer-ticks. */
121        tx_thread_sleep(4);
122
123        /* Put an instance in the counting semaphore. */
124        status = tx_semaphore_put (&my_semaphore);
125        if (status ! = TX_SUCCESS) break; /* Check status */
126
127        /* Activity 4: 3 timer-ticks. */
128        tx_thread_sleep(3);
129
130       /* Increment the thread counter and get timing info */
131       Speedy_Thread_counter + +;
132
133       current_time = tx_time_get();
134       cycle_time = current_time - start_time;
135       total_speedy_time = total_speedy_time + cycle_time;
136      }
137  }
138
139  /****************************************************/
140
141  /* Define the activities for the Consumer (Slow) thread. */
142
143  void Slow_Thread_entry(ULONG thread_input)
144  {
145
146  UINT status;
147  ULONG start_time, current_time, cycle_time;
148
149     while(1)
150     {
151       /* Get the starting time for this cycle */
152       start_time = tx_time_get();
153
154       /* Activity 5 - get an instance of the counting semaphore
155       with suspension and hold it for 12 timer-ticks. */
156
157       status = tx_semaphore_get (&my_semaphore, TX_WAIT_FOREVER);
158       if (status ! = TX_SUCCESS) break; /* Check status */
159
```

```
160       tx_thread_sleep(12);
161
162       /* Activity 6: 8 timer-ticks. */
163       tx_thread_sleep(8);
164
165
166       /* Activity 7: get an instance of the counting semaphore
167       with suspension and hold it for 11 timer-ticks. */
168
169       status = tx_semaphore_get (&my_semaphore, TX_WAIT_FOREVER);
170
171
172       if (status ! = TX_SUCCESS) break; /* Check status */
173
174       tx_thread_sleep(11);
175
176       /* Activity 8: 9 timer-ticks. */
177       tx_thread_sleep(9);
178
179       /* Increment the thread counter and get timing info */
180       Slow_Thread_counter++;
181
182       current_time = tx_time_get();
183       cycle_time = current_time - start_time;
184       total_slow_time = total_slow_time + cycle_time;
185   }
186 }
187
188 /************************************************/
189
190     /* Display statistics at periodic intervals */
191
192     void print_stats (ULONG invalue)
193 {
194     ULONG current_time, avg_slow_time, avg_speedy_time;
195
196     if ((Speedy_Thread_counter > 0) &&
        (Slow_Thread_counter > 0))
197     {
198        current_time = tx_time_get();
```

```
199        avg_slow_time = total_slow_time / Slow_Thread_counter;
200        avg_speedy_time = total_speedy_time /
             Speedy_Thread_counter;
201
202    printf("\nProducer-Consumer System - Timing Summary\n");
203    printf(" Current Time:           %lu\n",
204          current_time);
205    printf(" Speedy_Thread counter:  %lu\n",
206          Speedy_Thread_counter);
207·   printf(" Speedy_Thread avg time: %lu\n",
208          avg_speedy_time);
209    printf(" Slow_Thread counter:    %lu\n",
210          Slow_Thread_counter);
211    printf(" Slow_Thread avg time:   %lu\n\n",
212          avg_slow_time);
213    }
214    else printf("Bypassing print_stats function, Current Time:
       %lu\n",
215    tx_time_get());
216  }
```

11.19 Counting Semaphore Internals

When the TX_SEMAPHORE data type is used to declare a counting semaphore, a Semaphore Control Block (SCB) is created, and that Control Block is added to a doubly linked circular list, as illustrated in Figure 11.22.

The pointer named tx_semaphore_created_ptr points to the first Control Block in the list. See the fields in the SCB for timer attributes, values, and other pointers.

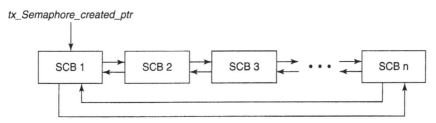

Figure 11.22: Created counting semaphore list

11.20 Overview

Both counting semaphores and mutexes can be used to provide mutual exclusion. However, mutexes should be used *only* for mutual exclusion, while counting semaphores are more versatile because they can also be used for event notification and thread synchronization.

A mutex is exceptionally robust in providing mutual exclusion. If this is crucial to your application, then using a mutex is a good decision. However, if mutual exclusion is not a major factor in your application, then use a counting semaphore because it is slightly faster and uses fewer system resources.

A special case of the counting semaphore is the binary semaphore, which has count values restricted to zero and one. If you want to use a counting semaphore for mutual exclusion, then you must use a binary semaphore.

Incrementing a semaphore's count is equivalent to placing an instance in the counting semaphore. Decrementing the count value corresponds to getting an instance from the counting semaphore.

There is no concept of ownership of counting semaphores as there is for mutexes. The producer-consumer system presented in this chapter illustrates this difference. Speedy_Thread placed instances in the semaphore without gaining ownership; Slow_Thread took instances from the semaphore whenever they were available, also without first gaining ownership. Furthermore, the priority inheritance feature is not available for counting semaphores as it is for mutexes. For a comparison of mutexes and counting semaphores, as well as recommended uses for each, refer to Figure 11.11 earlier in the chapter.

11.21 Key Terms and Phrases

binary semaphore	instance
Control Block	mutual exclusion
counting semaphore	place an instance
creating a semaphore	priority inversion
current count	producer-consumer system
deadlock	put operation
deadly embrace	retrieve an instance

decrement count	semaphore
deleting a semaphore	semaphore ceiling
event notification	semaphore information retrieval
FIFO order	suspend on semaphore
get operation	suspension list
increment count	synchronization

11.22 Problems

1. Describe a scenario in which you would use a binary semaphore rather than a mutex.

2. Assume that you have a common resource that can be shared by no more than three threads. Describe how you would use a counting semaphore to handle this situation.

3. Discuss why the count of a binary semaphore is usually initialized to one when it is created.

4. Describe how you would modify the producer-consumer system discussed in this chapter so that the current count of the semaphore would be displayed by the print_ stats expiration function.

5. What would happen if a thread placed an instance in a counting semaphore where the current count equaled 0xFFFFFFFF?

6. Describe what you would do to stop a thread from placing an instance in a counting semaphore that had a current count equal to 0xFFFFFFFF.

7. What would happen if the tx_semaphore_prioritize service was invoked, but no thread was in a suspended state?

8. Assume that my_semaphore has been declared as a counting semaphore. Describe two possible outcomes of invoking the following: tx_semaphore_get (&my_ semaphore, 5); (Hint: Check Appendix G.)

CHAPTER 12

Synchronization of Threads Using Event Flags Groups

12.1 Introduction

Event flags provide a powerful tool for thread synchronization. Event flags can be set or cleared[1] by any thread and can be inspected by any thread. Threads can suspend while waiting for some combination of event flags to be set. Each event flag is represented by a single bit. Event flags are arranged in groups of 32 as illustrated by Figure 12.1.

Threads can operate on all 32 event flags in a group simultaneously. To set or clear event flags, you use the tx_event_flags_set service and you "get" them (wait on them) with the tx_event_flags_get service.

Setting or clearing event flags is performed with a logical AND or OR operation between the current event flags and the new event flags. The user specifies the type of logical operation (either AND or OR) in the call to the tx_event_flags_set service.

There are similar logical options for getting event flags. A get request can specify that all specified event flags are required (a logical AND). Alternatively, a get request can specify that any of the specified event flags will satisfy the request (a logical OR). The user specifies the type of logical operation in the tx_event_flags_get call.

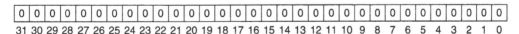

Figure 12.1: An event flags group

---

[1]We set a flag by storing the value 1 in that flag. We clear a flag by storing the value 0 in that flag.

Event flags that satisfy a get request are cleared if the request specifies either of the options TX_OR_CLEAR or TX_AND_CLEAR. The flag values remain unchanged when you use the TX_AND or TX_OR options in a get request.

Each event flags group is a public resource. ThreadX imposes no constraints as to how an event flags group can be used.

An application can create event flags groups either during initialization or during run-time. At the time of their creation, all event flags in the group are initialized to zero. There is no limit to the number of event flags groups an application may use.

Application threads can suspend while attempting to get any logical combination of event flags from a group. Immediately after one or more flags of a group have been set, ThreadX reviews the get requests of all threads suspended on that event flags group.[2] All the threads whose get requests were satisfied by the set operation are resumed.

As noted above, when at least one flag of a group has been set, ThreadX reviews all the threads suspended on that group. This review process creates overhead, so try to limit the number of threads using the same group to a reasonable number.

12.2 Event Flags Group Control Block

The characteristics of each event flags group are found in its Control Block. It contains information such as the values of current event flags, the reset search flag, the pointer to the suspension list for this event flags group, and the number of threads suspended for this group.[3] Figure 12.2 contains many of the fields that comprise this Control Block.

An Event Flags Group Control Block (ECB) can be located anywhere in memory, but it is common to make the Control Block a global structure by defining it outside the scope of any function. An ECB is created when an event flags group is declared with the TX_EVENT_FLAGS data type. For example, we declare my_event_group as follows:

```
TX_EVENT_FLAGS_GROUP my_event_group;
```

---

[2]More precisely, if the TX_OR option is used with the tx_event_flags_set service, then a review of the suspension list will occur. If the TX_AND option is used, no such review will be performed. See section 12.8 for more information.

[3]The structure of the Event Flags Group Control Block is defined in the **tx_api.h** file.

Field	Description
tx_event_flags_id	Event Flags Group ID
tx_event_flags_name	Pointer to event flags group name
tx_event_flags_current	Actual current event flags in this group
tx_event_flags_reset_search	Reset search flag set when ISR sets flags during search of suspended threads list
tx_event_flags_suspension_list	Pointer to event flags group suspension list
tx_event_flags_suspended_count	Number of threads suspended for event flags group
tx_event_flags_created_next	Pointer to next event flags group in the created list
tx_event_flags_created_previous	Pointer to previous event flags group in created list

Figure 12.2: Event Flags Group Control Block

The declaration of event flags groups normally appears in the declaration and definition section of the program.

12.3 Summary of Event Flags Group Control Services

Appendix C contains detailed information about event flags group services. This appendix contains information about each service, such as the prototype, a brief description of the service, required parameters, return values, notes and warnings, allowable invocation, and an example showing how the service can be used.

Figure 12.3 contains a list of all available services for an event flags group. We will investigate each of these services in the subsequent sections of this chapter.

12.4 Creating an Event Flags Group

An event flags group is declared with the TX_EVENT_FLAGS_GROUP data type and is defined with the tx_event_flags_create service. When defining an event flags group, you must specify its Control Block and the name of the event flags group. When created, all the event flags of a group are initialized to zero. Figure 12.4 lists the attributes of an event flags group.

Figure 12.5 illustrates how to use this service to create an event flags group. We will give our event flags group the name "my_event_group." If the variable *status* contains the return value TX_SUCCESS, we have successfully created an event flags group.

Event flags group service	Description
tx_event_flags_create	Create an event flags group
tx_event_flags_delete	Delete an event flags group
tx_event_flags_get	Get event flags from an event flags group
tx_event_flags_info_get	Retrieve information about an event flags group
tx_event_flags_performance info_get	Get event flags group performance information
tx_event_flags_performance_system_info_get	Retrieve performance system information
tx_event_flags_set	Set event flags in an event flags group
tx_event_flags_set_notify	Notify application when event flags are set

Figure 12.3: Services of the event flags group

Event flags group control block
Event flags group name
Group of 32 one-bit event flags

Figure 12.4: Attributes of an event flags group

```
TX_EVENT_FLAGS_GROUP my_event_group;
UINT status;

/* Create an event flags group. */

status = tx_event_flags_create(&my_event_group,
                             "my_event_group_name");

/* If status equals TX_SUCCESS, my_event_group is ready
   for get and set services. */
```

Figure 12.5: Creating an event flags group

12.5 Deleting an Event Flags Group

The tx_event_flags_delete service deletes an event flags group. When a group is deleted, all threads suspended on it resume and receive a TX_DELETED return status. Make certain that you do not attempt to use an event flags group that has been deleted. Figure 12.6 shows how to delete an event flags group.

If variable *status* contains the return value TX_SUCCESS, we have successfully deleted the event flags group.

12.6 Getting Event Flags from an Event Flags Group

The tx_event_flags_get service "gets," or waits on event flags from an event flags group. A get request is *satisfied* if the requested flags have been set in the specified event flags

```
TX_EVENT_FLAGS_GROUP my_event_group;
UINT status;

...

/* Delete event flags group. Assume that the group has
   already been created with a call to
   tx_event_flags_create. */

status = tx_event_flags_delete(&my_event_group);

/* If status equals TX_SUCCESS, the event flags group
   has been deleted. */
```

Figure 12.6: Deleting an event flags group

Get option	Description
TX_AND	All requested event flags must be set in the specified event flags group
TX_AND_CLEAR	All requested event flags must be set in the specified event flags group; event flags that satisfy the request are cleared
TX_OR	At least one requested event flag must be set in the specified event flags group
TX_OR_CLEAR	At least one requested event flag must be set in the specified event flags group; event flags that satisfy the request are cleared

Figure 12.7: Options to satisfy a get request

0	0	0	0	0	0	0	0	0	0	0	0	0	0	0	0	0	0	0	0	0	0	0	1	0	0	0	1	0	0	0	1
31	30	29	28	27	26	25	24	23	22	21	20	19	18	17	16	15	14	13	12	11	10	9	8	7	6	5	4	3	2	1	0

Figure 12.8: Example of an event flags group in which flags 0, 4, and 8 are set

```
TX_EVENT_FLAGS_GROUP my_event_group;
ULONG actual_events;
UINT status;

...

/* Retrieve event flags 0, 4, and 8 if they are all set.
   Also,if they are set they will be cleared. If the event
   flags are not set, this service suspends for a maximum
   of 20 timer-ticks. */

status = tx_event_flags_get(&my_event_group, 0x111,
                            TX_AND_CLEAR,
                            &actual_events, 20);

/* If status equals TX_SUCCESS, actual_events contains the
   actual events obtained, and event flags 0, 4, and 8 have
   been cleared from the event flags group. */
```

Figure 12.9: Getting event flags from an event flags group

group. The *wait_option* determines what action will be taken if the get request is not satisfied. The process of satisfying a get request depends on the *get_option,* which is a logical AND or OR operation, as depicted in Figure 12.7.

For example, assume that we want to determine whether event flags 0, 4, and 8 are all set. Furthermore, if those flags are all set, then we want them all cleared. Figure 12.8 illustrates an event flags group with flags 0, 4, and 8 set.

This corresponds to a hexadecimal value of 0x111, which we will use in our sample get operation. If all the desired event flags are not set when the request is made, then we will specify a maximum wait of 20 timer-ticks for them to become set. We will use the tx_event_flags_get service in Figure 12.9 in an attempt to determine whether event flags 0, 4, and 8 are all set.

If return variable *status* equals TX_SUCCESS, then event flags 0, 4, and 8 were found in a set state, and those flags were subsequently cleared. The variable *actual_events* contains the state of those flags as found before they were cleared, as well as the state of the remaining flags from the event flags group.

0	0	0	0	0	0	0	0	0	0	0	0	0	0	0	0	0	0	0	0	0	0	1	0	1	0	0	1	1	0	1	1	1
31	30	29	28	27	26	25	24	23	22	21	20	19	18	17	16	15	14	13	12	11	10	9	8	7	6	5	4	3	2	1	0	

Figure 12.10: Event flags group with a value of 0x537

0	0	0	0	0	0	0	0	0	0	0	0	0	0	0	0	0	0	0	0	0	1	0	0	0	0	1	0	0	1	1	0
31	30	29	28	27	26	25	24	23	22	21	20	19	18	17	16	15	14	13	12	11	10	9	8	7	6	5	4	3	2	1	0

Figure 12.11: Event flags group with a value of 0x426

0	0	0	0	0	0	0	0	0	0	0	0	0	0	0	0	0	0	0	0	0	1	0	0	0	0	1	0	0	0	0	1
31	30	29	28	27	26	25	24	23	22	21	20	19	18	17	16	15	14	13	12	11	10	9	8	7	6	5	4	3	2	1	0

Figure 12.12: Event flags group with flags 0, 5, and 10 set

Suppose that the value of the event flags group is 0x537, as represented in Figure 12.10 and the tx_event_flags_get service in Figure 12.9 is invoked.

After calling this service, the get operation is satisfied and the new value of the event flags group is 0x426, as illustrated in Figure 12.11.

The variable *actual_events* now contains 0x537, which is the original value of the event flags group. By contrast, if the *get_option* was TX_AND (rather than TX_AND_CLEAR), the get operation would have also been satisfied and the event flags group would have remained unchanged with the value 0x537. In this case, the variable *actual_events* would have also contained the value 0x537 after the service returned.

The previous example uses the TX_AND and the TX_AND_CLEAR *get_options*. We will consider another example that illustrates the effect of the TX_OR and the TX_OR_CLEAR *get_options*. Assume that we want to determine whether at least one of the event flags 0, 5, and 10 is set. Furthermore, we will clear all those flags that are set. Figure 12.12 illustrates an event flags group with flags 0, 5, and 10 set.

This corresponds to a hexadecimal value of 0x421, which we will use in our get operation. If none of the event flags are set when we make the get request, then we will wait indefinitely for at least one of them to become set. Assume that the value of the event flags group is 0x537. We will use the TX_OR *get_option* with the tx_event_flags_get service in Figure 12.13 to determine whether one or more of event flags 0, 5, and 10 is set.

```
TX_EVENT_FLAGS_GROUP my_event_group;
ULONG actual_events;
UINT status;

...

/* Retrieve event flags 0, 5, and 10 if at least one is set.
   If none of the event flags is set, this service suspends
   indefinitely waiting for at least one event flag to be set.
*/

status = tx_event_flags_get(&my_event_group, 0x421,
                            TX_OR, &actual_events,
                            TX_WAIT_FOREVER);

/* If status equals TX_SUCCESS, at least one flag was set.
   The event flags group remains unchanged and actual_events
   contains a copy of the event flags group value */
```

Figure 12.13: Getting event flags from an event flags group

0	0	0	0	0	0	0	0	0	0	0	0	0	0	0	0	0	1	1	1	1	1	1	1	1	0	0	0	0	1	1	0	0
31	30	29	28	27	26	25	24	23	22	21	20	19	18	17	16	15	14	13	12	11	10	9	8	7	6	5	4	3	2	1	0

Figure 12.14: Event flags group with value 0xFF0C

0	0	0	0	0	0	0	0	0	0	0	0	0	0	0	0	0	1	1	1	1	1	1	0	1	1	0	0	0	0	1	1	0	0
31	30	29	28	27	26	25	24	23	22	21	20	19	18	17	16	15	14	13	12	11	10	9	8	7	6	5	4	3	2	1	0

Figure 12.15: Event flags group with value 0xFB0C

If return variable *status* equals TX_SUCCESS, at least one of event flags 0, 5, and 10 was set and the value of the event flags group remains unchanged. The variable *actual_events* contains the value 0x537, which is the original value of the event flags group.

Suppose that the value of the event flags group is 0xFF0C, as represented in Figure 12.14 and the tx_event_flags_get service in Figure 12.13 is invoked.

After calling this service, the get operation is satisfied because flag 10 is set. The value of the event flags group remains unchanged, and the variable *actual_events* contains a copy of this value. However, if we used the TX_OR_CLEAR *get_option* then the event flags group would change to a value of 0xFB0C, as represented in Figure 12.15. (Flag 10 would be cleared by the get operation.)

The return variable *actual_events* would contain 0xFF0C, which was the original value of the event flags group.

12.7 Retrieving Information about an Event Flags Group

There are three services that enable you to retrieve vital information about event flags groups. The first such service for event flags groups—the tx_event_flags_info_get service—retrieves a subset of information from the Event Flags Group Control Block. This information provides a "snapshot" at a particular instant in time, i.e., when the service is invoked. The other two services provide summary information that is based on the gathering of run-time performance data. One service—the tx_event_flags_performance_info_get service—provides an information summary for a particular event flags group up to the time the service is invoked. By contrast the tx_event_flags_performance_system_info_get retrieves an information summary for all event flags groups in the system up to the time the service is invoked. These services are useful in analyzing the behavior of the system and determining whether there are potential problem

```
TX_EVENT_FLAGS_GROUP my_event_group;
CHAR *name;
ULONG current_flags;
TX_THREAD *first_suspended;
ULONG suspended_count;
TX_EVENT_FLAGS_GROUP *next_group;
UINT status;

…

/* Retrieve information about the previously created
   event flags group named "my_event_group." */

status = tx_event_flags_info_get(&my_event_group, &name,
                                 &current_flags,
                                 &first_suspended,
                                 &suspended_count,
                                 &next_group);

/* If status equals TX_SUCCESS, the information requested
   is valid. */
```

Figure 12.16: Retrieving information about an event flags group

areas. The tx_event_flags_info_get[4] service retrieves several useful items of information about an event flags group. This information includes the name of the event flags group, the current value of the event flags, the number of threads suspended for this group, and a pointer to the next created event flags group. Figure 12.16 shows how this service can be used to obtain information about an event flags group.

If return variable *status* contain the value TX_SUCCESS, we have retrieved valid information about the event flags group.

12.8 Setting Event Flags in an Event Flags Group

The tx_event_flags_set service sets or clears one or more event flags in a group. When the set service is performed and actually sets one or more of the flags, the scheduler checks whether there are any threads suspended for that event flags group. If there are threads suspended for the resulting value of this group, then those threads are resumed.

The process of setting or clearing event flags depends on the *set_option*, which is a logical AND or OR operation, as depicted in Figure 12.17.[5]

For example, suppose that we want to clear all flags except flags 0, 4, and 8, in which the current values of the event flags group is 0xFF0C. We would pass the value 0x111 (i.e., event flags 0, 4, and 8) and use the TX_AND option. Figure 12.18 illustrates this operation.

Set option	Description
TX_AND	The specified event flags are ANDed into the current event flags group; this option is often used to clear event flags in a group
TX_OR	The specified event flags are ORed with the current event flags group[6]

Figure 12.17: Set options

---

[4]By default, only the *tx_event_flags_info_get* service is enabled. The other two information gathering services must be enabled in order to use them.

[5]The TX_OR option forces the scheduler to review the suspension list to determine whether any threads are suspended for this event flags group.

[6]The TX_OR option forces the scheduler to review the suspension list to determine whether any threads are suspended for this event flags group.

The new value of the event flags group is 0x100 because flag 8 is the only flag in common for the values 0xFF0C and 0x112. However, if the TX_OR option is used, then the new event flags group value is 0xFF1D, as illustrated in Figure 12.19.

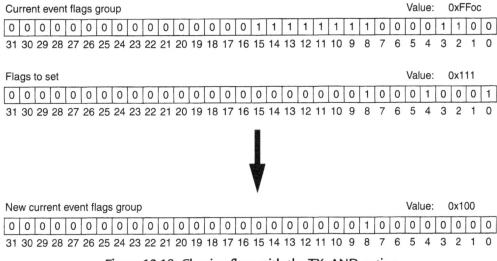

Figure 12.18: Clearing flags with the TX_AND option

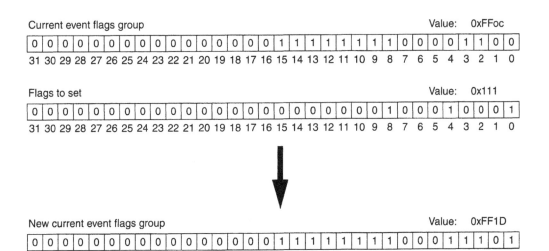

Figure 12.19: Setting flags with the TX_OR option

```
TX_EVENT_FLAGS_GROUP my_event_group;
UINT status;

/* Set event flags 0, 4, and 8. */

...

status = tx_event_flags_set(&my_event_group,
                            0x111, TX_OR);

/* If status equals TX_SUCCESS, the event flags have been
   set and any suspended thread whose request was satisfied
   has been resumed. */
```

Figure 12.20: Set event flags in an event flags group

Figure 12.20 illustrates how the tx_event_flags_set service can be used to set the value 0x111 with the TX_OR option.

If return variable *status* contains the value of TX_SUCCESS, the requested event flags have been set. Any threads that were suspended on these event flags have been resumed.

12.9 Event Flags Group Notification and Event-Chaining

The tx_event_flags_set_notify service registers a notification callback function that is invoked whenever one or more event flags are set in the specified event flags group. The processing of the notification callback is defined by the application. This is an example of event-chaining where notification services are used to chain various synchronization events together. This is typically useful when a single thread must process multiple synchronization events.

12.10 Sample System Using an Event Flags Group to Synchronize Two Threads

We used counting semaphores for mutual exclusion and for event notification in the two previous sample systems. In this sample system, we will focus on synchronizing thread behavior by using an event flags group. An event flags group provides a powerful means of communication between threads because it consists of 32 one-bit flags, thus providing an extensive number of flag combinations. We will modify the previous sample system and replace all references to a counting semaphore with references to an event flags group.

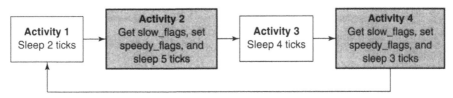

Figure 12.21: Activities of the Speedy_Thread

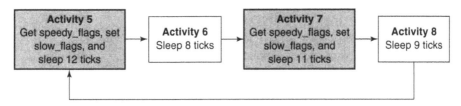

Figure 12.22: Activities of the Slow_Thread where priority = 15

We will restrict our attention to two flag values that will be used by Speedy_Thread and Slow_Thread to communicate with each other. There are no critical sections in this sample system. The threads synchronize their activities with the event flags group. The two event flag values we will use are 0xF0, which we will call *speedy_flags,* and 0x0F, which we will call *slow_flags.*[7] In Figure 12.21, when Speedy_Thread enters Activity 2 or Activity 4, it tries to get slow_flags. If it is successful, it clears those flags, sets speedy_flags, and continues processing.

In Figure 12.22, when Slow_Thread enters Activity 5 or Activity 7, it tries to get speedy_flags. If it is successful, it clears those flags, sets slow_flags, and continues processing.

Following is a portion of the output produced by this sample system.

Event Flags Group synchronizes 2 threads	
Current Time:	500
Speedy_Thread counter:	13
Speedy_Thread avg time:	37
Slow_Thread counter:	12
Slow_Thread avg time:	40

---

[7]Note that speedy_flags and slow_flags each represent 4 requested event flags set out of a possible 32. We use these constants to illustrate the use of TX_AND and TX_OR operations.

Event Flags Group synchronizes 2 threads
Current Time: 1000
Speedy_Thread counter: 25
Speedy_Thread avg time: 38
Slow_Thread counter: 24
Slow_Thread avg time: 40

Note that Speedy_Thread and Slow_Thread each complete about the same number of cycles, even though Speedy_Thread has a higher priority. In this sample system, the event flags group serves as a toggle switch and ensures that the threads take an equal number of turns processing.

We will discuss a series of changes to be applied to the system from Chapter 10 so that all references to a binary semaphore will be replaced with references to an event flags group. The complete program listing, called *12_sample_system.c,* is located in the next section of this chapter and on the attached CD.

The first change occurs in the declaration and definitions section of our program, in which we replace the declaration of a counting semaphore with the declaration of an event flags group, as follows.

```
TX_EVENT_FLAGS_GROUP  my_event_group;
```

We also declare and initialize the variables *speedy_flags, slow_flags,* and *actual_flags* as follows:

```
ULONG     speedy_flags = 0xF0,
          slow_flags = 0x0F
          actual_events;
```

We will use the variable *actual_flags* for our get operations. The next change occurs in the application definitions section of our program, in which we replace the creation of a binary semaphore with the creation and initialization of an event flags group, as follows.

```
/* Create the event flags group used by both threads, initialize to
   slow_flags (0X0F). */
tx_event_flags_create (&my_event_group, "my_event_group");
tx_event_flags_set (&my_event_group, slow_flags, TX_OR);
```

```
/* Activity 2 - Wait for slow_flags in the event flags group,
   set it to speedy_flags, and hold it for 5 timer-ticks.  */

status = tx_event_flags_get (&my_event_group,
                             slow_flags, TX_AND_CLEAR,
                             &actual_events, TX_WAIT_FOREVER);
if (status != TX_SUCCESS) break;    /* Check status.  */

status = tx_event_flags_set (&my_event_group,
                             speedy_flags, TX_OR);
if (status != TX_SUCCESS) break;    /* Check status.  */

tx_thread_sleep(5);
```

Figure 12.23: Changes to Activity 2


```
/* Activity 4 - Wait for slow_flags in the event flags group,
   set it to speedy_flags, and hold it for 3 timer-ticks.  */

status = tx_event_flags_get (&my_event_group,
                             slow_flags, TX_AND_CLEAR,
                             &actual_events, TX_WAIT_FOREVER);
if (status != TX_SUCCESS) break;    /* Check status.  */

status = tx_event_flags_set (&my_event_group,
                             speedy_flags, TX_OR);
if (status != TX_SUCCESS) break;    /* Check status.  */
```

Figure 12.24: Changes to Activity 4

We arbitrarily set the event flags group to the value *slow_flags* in the preceding statement. The only consequence of this particular initialization is that Speedy_Thread will be the first thread to execute. We could have set the event flags group to the value *speedy_flags*, thereby giving Slow_Thread the first opportunity to execute.

The remaining changes occur in the function definitions section of our program. We need to change all references to a binary semaphore with references to an event flags group. We will show only the changes for the Speedy_Thread and will leave the Slow_Thread changes as an exercise for the reader. Figure 12.23 contains the necessary changes for Activity 2.

Figure 12.24 contains the necessary changes for Activity 4. Most of the modifications involve changing binary semaphore calls to event flags group calls.

12.11 Listing for 12_sample_system.c

The sample system named *12_sample_system.c* is located on the attached CD. The complete listing appears below; line numbers have been added for easy reference.

```
001   /* 12_sample_system.c

002
003      Create two threads and one event flags group.
004      The threads synchronize their behavior via the
005      event flags group. */
006

007
008   /****************************************************/
009   /*      Declarations, Definitions, and Prototypes     */
010   /****************************************************/

011
012   #include    "tx_api.h"
013   #include    <stdio.h>

014
015   #define      STACK_SIZE      1024
016

017   /* Declare stacks for both threads. */
018   CHAR stack_speedy[STACK_SIZE];
019   CHAR stack_slow[STACK_SIZE];
020

021   /* Define the ThreadX object control blocks. */
022   TX_THREAD        Speedy_Thread;
023   TX_THREAD        Slow_Thread;
024   TX_EVENT_FLAGS_GROUP my_event_group;
025

026   /* Declare the application timer */
027   TX_TIMER stats_timer;
028
```

```
029   /* Declare the counters, accumulators, and flags */
030   ULONG     Speedy_Thread_counter = 0,
031             total_speedy_time = 0;
032   ULONG     Slow_Thread_counter = 0,
033             total_slow_time = 0;
034   ULONG     slow_flags = 0X0F,
035             speedy_flags = 0XF0,
036             actual_events;
037
038   /* Define thread prototypes. */
039   void Speedy_Thread_entry(ULONG thread_input);
040   void Slow_Thread_entry(ULONG thread_input);
041
042   /* Define prototype for expiration function */
043   void print_stats(ULONG);
044
045
046   /******************************************************/
047   /*                 Main Entry Point                  */
048   /******************************************************/
049
050   /* Define main entry point. */
051
052   int main()
053   {
054      /* Enter the ThreadX kernel. */
055      tx_kernel_enter();
056   }
057
058
059   /******************************************************/
060   /*              Application Definitions              */
061   /******************************************************/
062
063   /* Define what the initial system looks like. */
064
065   void tx_application_define(void *first_unused_memory)
066   {
067      /* Put system definitions here,
068          e.g., thread and event flags group creates */
```

```
069
070        /* Create the Speedy_Thread. */
071        tx_thread_create(&Speedy_Thread, "Speedy_Thread",
072                        Speedy_Thread_entry, 0,
073                        stack_speedy, STACK_SIZE,
074                        5, 5, TX_NO_TIME_SLICE, TX_AUTO_START);
075
076        /* Create the Slow_Thread */
077        tx_thread_create(&Slow_Thread, "Slow_Thread",
078                        Slow_Thread_entry, 1,
079                        stack_slow, STACK_SIZE,
080                         15, 15, TX_NO_TIME_SLICE, TX_AUTO_START);
081
082        /* Create the event flags group used by both threads,
083           initialize to slow_flags (0X0F). */
084        tx_event_flags_create (&my_event_group, "my_event_group");
085        tx_event_flags_set (&my_event_group, slow_flags, TX_OR);
086
087        /* Create and activate the timer */
088        tx_timer_create (&stats_timer, "stats_timer", print_stats,
089                        0x1234, 500, 500, TX_AUTO_ACTIVATE);
090
091    }
092
093
094    /****************************************************/
095    /*                 Function Definitions             */
096    /****************************************************/
097
098    /* "Speedy_Thread" - it has a higher priority than the other
       thread */
099
100    void Speedy_Thread_entry(ULONG thread_input)
101    {
102
103    UINT      status;
104    ULONG     start_time, cycle_time, current_time;
105
106        while(1)
107        {
```

```
108          /* Get the starting time for this cycle */
109          start_time = tx_time_get();
110
111          /* Activity 1: 2 timer-ticks. */
112          tx_thread_sleep(2);
113
114          /* Activity 2 - Wait for slow_flags in the event
             flags group, set it to speedy_flags, and hold it for
             5 timer-ticks. */
115
116
117          status = tx_event_flags_get (&my_event_group, slow_
                flags, TX_AND_CLEAR,
118                                          &actual_events, TX_WAIT_
                                             FOREVER);
119          if (status ! = TX_SUCCESS) break; /* Check status. */
120
121          status = tx_event_flags_set (&my_event_group, speedy_
                flags, TX_OR);
122          if (status ! = TX_SUCCESS) break; /* Check status. */
123
124          tx_thread_sleep(5);
125
126          /* Activity 3: 4 timer-ticks. */
127          tx_thread_sleep(4);
128
129          /* Activity 4 - Wait for slow_flags in the event
             flags group, set it to speedy_flags, and hold it for
             3 timer-ticks. */
130
131
132          status = tx_event_flags_get (&my_event_group, slow_flags,
             TX_AND_CLEAR,
133                                          &actual_events,
                                             TX_WAIT_FOREVER);
134
135          if (status ! = TX_SUCCESS) break; /* Check status. */
136
137          status = tx_event_flags_set (&my_event_group, speedy_
             flags, TX_OR);
```

```
138          if (status ! = TX_SUCCESS) break; /* Check status. */
139
140          tx_thread_sleep(3);
141
142          /* Increment the thread counter and get timing info */
143          Speedy_Thread_counter + +;
144
145          current_time = tx_time_get();
146          cycle_time = current_time - start_time;
147          total_speedy_time = total_speedy_time + cycle_time;
148
149      }
150  }
151
152  /********************************************************/
153  /* "Slow_Thread" - it has a lower priority than the other
     thread */
154
155  void Slow_Thread_entry(ULONG thread_input)
156  {
157
158  UINT status;
159  ULONG start_time, current_time, cycle_time;
160
161      while(1)
162      {
163        /* Get the starting time for this cycle */
164        start_time = tx_time_get();
165
166        /* Activity 5 - Wait for speedy_flags in the event
           flags group, set it to slow_flags, and hold it for
           12 timer-ticks. */
167
168        status = tx_event_flags_get (&my_event_group, speedy_flags,
           TX_AND_CLEAR,
169                                   &actual_events,
                                      TX_WAIT_FOREVER);
170
171        if (status ! = TX_SUCCESS) break;  /* Check status. */
172
```

```
173        status = tx_event_flags_set (&my_event_group, slow_flags,
           TX_OR);
174        if (status ! = TX_SUCCESS) break;   /* Check status. */
175
176        tx_thread_sleep(12);
177
178        /* Activity 6: 8 timer-ticks. */
179        tx_thread_sleep(8);
180
181        /* Activity 7: Wait for speedy_flags in the event flags
           group, set it
182        to slow_flags, and hold it for 11 timer-ticks. */
183
184        status = tx_event_flags_get (&my_event_group, speedy_flags,
      TX_AND_CLEAR,
185                                      &actual_events,
                                        TX_WAIT_FOREVER);
186
187        if (status ! = TX_SUCCESS) break; /* Check status. */
188
189        status = tx_event_flags_set (&my_event_group, slow_flags,
      TX_OR);
190
191        tx_thread_sleep(11);
192
193        /* Activity 8: 9 timer-ticks. */
194        tx_thread_sleep(9);
195
196        /* Increment the thread counter and get timing info */
197        Slow_Thread_counter++;
198
199        current_time = tx_time_get();
200        cycle_time = current_time - start_time;
201        total_slow_time = total_slow_time + cycle_time;
202
203    }
204  }
205
206  /*****************************************************/
207  /* print statistics at specified times */
```

```
208   void print_stats (ULONG invalue)
209   {
210       ULONG current_time, avg_slow_time, avg_speedy_time;
211
212       if ((Speedy_Thread_counter>0) && (Slow_Thread_counter>0))
213         {
214       current_time = tx_time_get();
215       avg_slow_time = total_slow_time / Slow_Thread_counter;
216       avg_speedy_time = total_speedy_time /
          Speedy_Thread_counter;
217
218       printf("\nEvent Flags Group synchronizes 2 threads\n");
219       printf("              Current Time:              %lu\n",
220           current_time);
221       printf("              Speedy_Thread counter:     %lu\n",
222           Speedy_Thread_counter);
223       printf("              Speedy_Thread avg time:    %lu\n",
224           avg_speedy_time);
225       printf("              Slow_Thread counter:       %lu\n",
226           Slow_Thread_counter);
227       printf("              Slow_Thread avg time:      %lu\n\n",
228           avg_slow_time);
229         }
230       else printf("Bypassing print_stats function, Current
                  Time: %lu\n", tx_time_get());
231
232   }
```

12.12 Event Flags Group Internals

When the TX_EVENT_FLAGS data type is used to declare an event flags group, an ECB is created, and that Control Block is added to a doubly linked circular list, as illustrated in Figure 12.25.

When flags become set in an event flags group, ThreadX immediately reviews all threads that are suspended on that event flags group. This introduces some overhead, so limit the number of threads using the same event flags group to a reasonable number.

tx_event_flags_created_ptr

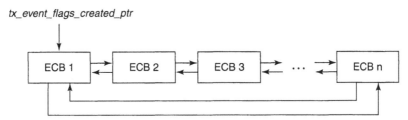

Figure 12.25: Created event flags group list

12.13 Overview

Event flags provide a powerful tool for thread synchronization. Event flags groups do not support a concept of ownership, nor is there a limit to how many threads can access an event flags group.

Event flags can be set by any thread and can be inspected by any thread.

Threads can suspend while waiting for some combination of event flags to be set.

Threads can operate on all 32 event flags in a group simultaneously. Threads can set or clear event flags using the tx_event_flags_set service and get them (wait on them) by using the tx_event_flags_get service.

The clearing or setting of event flags entails a logical TX_AND or TX_OR operation between the current event flags and the new event flags. There are similar logical options for getting event flags. A get request can specify that all specified event flags are required (a logical TX_AND). Alternatively, a get request can specify that any of the specified event flags will satisfy the request (a logical TX_OR).

Event flags that satisfy a get request are cleared if either of the *clear options* TX_ OR_ CLEAR or TX_AND_CLEAR are specified by the request. The event flag values remain unchanged when the TX_AND or TX_OR options are used in a get request.

Application threads can suspend while attempting to get any logical combination of event flags from a group. When at least one event flag becomes set, the get requests of all threads suspended on that event flags group are reviewed. All the threads whose get requests are now satisfied are resumed.

As noted above, when at least one flag of a group becomes set, ThreadX reviews all the threads suspended on that group. This review process creates overhead, so try to limit the number of threads using the same event flags group to a reasonable number.

12.14 Key Terms and Phrases

clearing event flags	initialization of event flags
creating an event flags group	logical operations
deleting an event flags group	retrieval of event flags
event flags group	satisfying a get request
Event Flags Group Control Block (ECB)	set option setting event flags
flag	suspension of threads
get option	synchronization of threads
get request	wait option

12.15 Problems

1. Compare mutexes, counting semaphores, and event flags groups. Describe three different scenarios in which using each object is better suited than using the other two objects.

2. Describe how you would determine how many threads are suspended for a certain event flags group.

3. Describe how you would determine the current value of an event flags group.

4. Suppose that you want to synchronize the operation of three threads. Describe how you would use an event flags group so that the threads are processed in a specific order, i.e., so that your application processes the first thread, the second thread, and then the third thread. (This process order is to repeat indefinitely.)

5. Suppose that you want to synchronize the operation of three threads. Describe how you would use an event flags group so that, at most, two of the threads can be processed at any time, but the third thread depends on one of the other two threads before it can execute.

6. Suppose that an event flags group contains one of the values 0x110, 0x101, or 0x011. Describe how you would perform a get operation that would be satisfied for any one of these values.

7. If an event flags group had the value 0xFDB, what successful get operation could have caused the value of this group to change to 0xBDA?

8. If an event flags group had the value 0xF4C, what successful set operation could have cause the value of this group to change to 0x148?

CHAPTER 13

Thread Communication with Message Queues

13.1 Introduction

Message queues are the primary means of interthread communication in ThreadX. One or more messages can reside in a message queue, which generally observes a FIFO discipline. A message queue that can hold just a single message is commonly called a *mailbox*.

The tx_queue_send service places messages in the rear of a queue and the tx_queue_receive service removes messages from the front of a queue. The only exception to this protocol occurs when a thread is suspended while waiting for a message from an empty queue. In this case, ThreadX places the next message sent to the queue directly into the thread's destination area, thus bypassing the queue altogether. Figure 13.1 illustrates a message queue.

Each message queue is a public resource. ThreadX places no constraints on how message queues are used.

Applications can create message queues either during initialization or during runtime. There is no limit to the number of message queues an application may use.

Figure 13.1: A message queue

Message queues can be created to support a variety of message sizes. The available message sizes are 1, 2, 4, 8, and 16 32-bit words. The message size is specified when the queue is created. If your application messages exceed 16 words, you must send your messages by pointer. To accomplish this, create a queue with a message size of one word (enough to hold a pointer), and then send and receive message pointers instead of the entire message.

The number of messages a queue can hold depends on its message size and the size of the memory area supplied during creation. To calculate the total message capacity of the queue, divide the number of bytes in each message into the total number of bytes in the supplied memory area.

For example, if you create a message queue that supports a message size of one 32-bit word (four bytes), and the queue has a 100-byte available memory area, its capacity is 25 messages.

The memory area for buffering messages is specified during queue creation. It can be located anywhere in the target's address space. This is an important feature because it gives the application considerable flexibility. For example, an application might locate the memory area of a very important queue in high-speed RAM to improve performance.

Application threads can suspend while attempting to send or receive a message from a queue. Typically, thread suspension involves waiting for a message from an empty queue. However, it is also possible for a thread to suspend trying to send a message to a full queue.

After the condition for suspension is resolved, the request completes and the waiting thread is resumed. If multiple threads are suspended on the same queue, they are resumed in the order they occur on the suspended list (usually FIFO). However, an application can cause a higher-priority thread to resume first by calling the tx_queue_prioritize service prior to the queue service that lifts thread suspension. The queue prioritize service places the highest-priority thread at the front of the suspension list, while leaving all other suspended threads in the same FIFO order.

Queue suspensions can also time out; essentially, a time-out specifies the maximum number of timer-ticks the thread will stay suspended. If a time-out occurs, the thread is resumed and the service returns with the appropriate error code.

13.2 Message Queue Control Block

The characteristics of each message queue are found in its Control Block. It contains information such as the message size, total message capacity, current number of messages in the queue, available queue storage space, and number of threads suspended on this message queue.[1] Figure 13.2 contains many of the fields that comprise this Control Block.

A Message Queue Control Block can be located anywhere in memory, but it is common to make the Control Block a global structure by defining it outside the scope of any function. A Message Queue Group Control Block is created when a message queue is declared with the TX_QUEUE data type. For example, we declare my_queue as follows:

```
TX_QUEUE my_queue;
```

The declaration of a message queue normally appears in the declarations and definitions section of the program.

Field	Description
tx_queue_id	Message Queue ID
tx_queue_name	Pointer to message queue name
tx_queue_message_size	Message size specified during queue creation
tx_queue_capacity	Total number of messages in the queue
tx_queue_enqueued	Current number of messages in the message queue
tx_queue_available_storage	Available message queue storage space
tx_queue_start	Pointer to the start of the queue message area
x_queue_end	Pointer to the end of the queue message area
tx_queue_read	Read pointer—used by receive requests
tx_queue_write	Write pointer—used by send requests
tx_queue_suspension_list	Pointer to the head of the queue suspension list
tx_queue_suspended_count	Count of how many threads are suspended
tx_queue_created_next	Pointer to next message queue in the created list
tx_queue_created_previous	Pointer to previous message queue in created list

Figure 13.2: Message Queue Control Block

---

[1]The structure of the Message Queue Control Block is defined in the **tx_api.h** file.

13.3 Summary of Message Queue Services

Appendix F contains detailed information about message queue services. This appendix contains information about each service, such as the prototype, a brief description of the service, required parameters, return values, notes and warnings, allowable invocation, and an example showing how the service can be used. Figure 13.3 contains a listing of all available services for a message queue. We will investigate each of these services in the subsequent sections of this chapter.

13.4 Creating a Message Queue

A message queue is declared with the TX_QUEUE data type and is defined with the tx_queue_create service. When defining a message queue, you must specify its Control Block, the name of the message queue, the message size, the starting address of the

Message Queue Service	Description
tx_queue_create	Create a message queue
tx_queue_delete	Delete a message queue
tx_queue_flush	Empty all messages in a message queue
tx_queue_front_send	Send a message to the front of a message queue
tx_queue_info_get	Retrieve information about a message queue
tx_queue_performance_info_get	Get queue performance information
tx_queue_performance_system_info_get	Get queue system performance information
tx_queue_prioritize	Prioritize a message queue suspension list
tx_queue_receive	Get a message from a message queue
tx_queue_send	Send a message to a message queue
tx_queue_send_notify	Notify application when message is sent to queue

Figure 13.3: Services of the message queue

queue, and the total number of bytes available for the message queue. Figure 13.4 lists the attributes of a message queue. The total number of messages is calculated from the specified message size and the total number of bytes in the queue. Note that if the total number of bytes specified in the queue's memory area is not evenly divisible by the specified message size, the remaining bytes in the memory area are not used.

Figure 13.5 illustrates the use of this service. We will give our message queue the name "my_queue."

13.5 Sending a Message to a Message Queue

The tx_queue_send service sends a message to the specified message queue. This service copies the message to be sent to the back of the queue from the memory area specified by

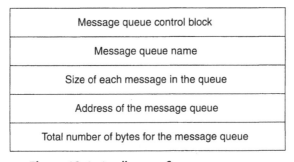

Figure 13.4: Attributes of a message queue

```
TX_QUEUE my_queue;
UINT status;

/* Create a message queue whose total size is 2000 bytes
   starting at address 0x300000. Each message in this
   queue is defined to be 4 32-bit words long. */

status = tx_queue_create(&my_queue, "my_queue_name",
                  TX_4_ULONG, (VOID *) 0x300000, 2000);

/* If status equals TX_SUCCESS, my_queue contains space
   for storing 125 messages: (2000 bytes / 16 bytes per message). */
```

Figure 13.5: Creating a message queue

```
TX_QUEUE my_queue;
UINT status;
ULONG my_message[4];

...

/* Send a message to "my_queue." Return immediately,
   regardless of success. This wait option is used for
   calls from initialization, timers, and ISRs. */

status = tx_queue_send(&my_queue, my_message, TX_NO_WAIT);

/* If status equals TX_SUCCESS, the message has been sent
   to the queue. */
```

Figure 13.6: Send a message to a queue

the source pointer. Figure 13.6 shows how this service can be used to send a message to a queue.[2]

If return variable *status* contains the value TX_SUCCESS, we have successfully sent a message to the queue.

13.6 Receiving a Message from a Message Queue

The tx_queue_receive service retrieves a message from a message queue. This service copies the retrieved message from the front of the queue into the memory area specified by the destination pointer. That message is then removed from the queue. The specified destination memory area must be large enough to hold the message; i.e., the destination pointed to by destination_ptr must be at least as large as this queue's defined message size. Otherwise, memory corruption occurs in the memory area following the destination. Figure 13.7 shows how this service can be used to receive a message from a queue.[3]

---

[2]Note that TX_NO_WAIT is the only valid wait option if this service is called from a non-thread, such as an application timer, initialization routine, or ISR.

[3]Note that TX_NO_WAIT is the only valid wait option if this service is called from a non-thread, such as an application timer, initialization routine, or ISR.

```
TX_QUEUE my_queue;
UINT status;
ULONG my_message[4];

...

/* Retrieve a message from "my_queue." If the queue is
   empty, suspend until a message is present. Note that
   this suspension is only possible from application
   threads. */
status = tx_queue_receive(&my_queue, my_message,
                          TX_WAIT_FOREVER);

/* If status equals TX_SUCCESS, the message is in
   "my_message." */
```

Figure 13.7: Receive a message from a queue

```
TX_QUEUE my_queue;
UINT status;

...

/* Delete entire message queue. Assume that the
   queue has already been created with a call to
   tx_queue_create. */
status = tx_queue_delete(&my_queue);

/* If status equals TX_SUCCESS, the message queue
   has been deleted. */
```

Figure 13.8: Deleting a message queue

13.7 Deleting a Message Queue

The tx_queue_delete service deletes a message queue. All threads that are suspended waiting for a message from this queue are resumed and receive a TX_DELETED return status. Do not attempt to use a message queue that has been deleted. Also, make certain that you manage the memory area associated with a queue that has been deleted. Figure 13.8 contains an example showing how to delete a message queue.

If variable *status* contains the return value TX_SUCCESS, we have successfully deleted the message queue.

```
TX_QUEUE my_queue;
UINT status;

...

/* Delete all messages in the message queue.
   Assume that the queue has already been created
   with a call to tx_queue_create. */
status = tx_queue_flush(&my_queue);

/* If status equals TX_SUCCESS, the message queue
   is now empty. */
```

Figure 13.9: Flushing a message queue

13.8 Flushing the Contents of a Message Queue

The tx_queue_flush service deletes all messages that are stored in a message queue. In addition, if the queue is full and there are threads suspended because of trying to send messages to that queue, then all the messages of those suspended threads are discarded, and each suspended thread is resumed with a successful return status. If the queue is empty, this service does nothing.

Figure 13.9 illustrates the use of this service. We will give our message queue the name "my_queue."

If variable *status* contains the return value TX_SUCCESS, we have successfully emptied the message queue.

13.9 Sending a Message to the Front of a Message Queue

Normally, messages are sent to the rear of a queue, but applications can use the tx_queue_front_send service to send a message to the front location of the message queue. This service copies the message to the front of the queue from the memory area specified by the source pointer. Figure 13.10 illustrates the use of this service. We will give our message queue the name "my_queue" and we will use the wait option that returns immediately, regardless of whether or not the message was successfully placed at the front of the queue. If the variable *status* contains the return value TX_SUCCESS, we have successfully sent a message to the front of a queue.

```
TX_QUEUE my_queue;
UINT status;
ULONG my_message[TX_4_ULONG];

...

/* Send a message to the front of "my_queue." Return
   immediately, regardless of success. This wait
   option is used for calls from initialization,
   timers, and ISRs. */
status = tx_queue_front_send(&my_queue, my_message,
                             TX_NO_WAIT);

/* If status equals TX_SUCCESS, the message has been
   placed at the front of the specified queue. */
```

Figure 13.10: Sending a message to the front of a queue

13.10 Retrieving Message Queue Information

There are three services that enable you to retrieve vital information about message queues. The first such service for message queues—the tx_queue_info_get service—retrieves a subset of information from the Message Queue Control Block. This information provides a "snapshot" at a particular instant in time, i.e., when the service is invoked. The other two services provide summary information that is based on the gathering of run-time performance data. One service—the tx_queue_performance_info_get service—provides an information summary for a particular queue up to the time the service is invoked. By contrast the tx_queue_performance_system_info_get retrieves an information summary for all message queues in the system up to the time the service is invoked. These services are useful in analyzing the behavior of the system and determining whether there are potential problem areas. The tx_queue_info_get[4] service retrieves several useful items of information about a message queue. These include the name of the message queue, the number of messages currently in the queue, the queue's total available storage (in bytes), the number of threads suspended for this queue, a pointer to the first suspended thread, and a pointer to the next created message queue. Figure 13.11 shows how this service can be used.

---

[4]By default, only the *tx_queue_info_get* service is enabled. The other two information-gathering services must be enabled in order to use them.

```
TX_QUEUE my_queue;
CHAR *name;
ULONG enqueued;
TX_THREAD *first_suspended;
ULONG suspended_count;
ULONG available_storage;
TX_QUEUE *next_queue;
UINT status;
...

/* Retrieve information about the previously
   created message queue "my_queue." */

status = tx_queue_info_get(&my_queue, &name,
                           &enqueued, &available_storage,
                           &first_suspended, &suspended_count,
                           &next_queue);

/* If status equals TX_SUCCESS, the information requested
   is valid. */
```

Figure 13.11: Retrieving information about a message queue

Status of Queue	Effect of Prioritization
Queue is empty	The highest priority thread suspended for this queue will receive the next message placed on the queue
Queue is full	The highest priority thread suspended for this queue will send the next message to this queue when space becomes available

Figure 13.12: Effect of prioritizing a message queue suspension list

If return variable *status* contains the value TX_SUCCESS, we have retrieved valid information about the message queue.

13.11 Prioritizing a Message Queue Suspension List

The tx_queue_prioritize service places the highest priority thread suspended for a message queue at the front of the suspension list. This applies either to a thread waiting to receive a message from an empty queue, or to a thread waiting to send a message to a full queue, as described in Figure 13.12. All other threads remain in the same FIFO order in which they were suspended.

```
TX_QUEUE my_queue;
UINT status;

/* Depending on the queue status, this service ensures
   that the highest priority thread will either
   receive the next message placed on this queue, or
   will send the next message to the queue. */
status = tx_queue_prioritize(&my_queue);

/* If status equals TX_SUCCESS, the highest priority
   suspended thread is at the front of the list. If the
   suspended thread is waiting to receive a message,
   the next tx_queue_send or tx_queue_front_send call
   made to this queue will wake up this thread. If the
   suspended thread is waiting to send a message, the
   next tx_queue_receive call will wake up this thread. */
```

Figure 13.13: Prioritizing a message queue suspension list

Figure 13.13 contains an example showing how this service can be used to prioritize a message queue suspension list.

If return variable *status* contains the value TX_SUCCESS, we have successfully prioritized the message queue suspension list.

13.12 Message Queue Notification and Event-Chaining

The tx_queue_send_notify service registers a notification callback function that is invoked whenever a message is sent to the specified queue. The processing of the notification callback is defined by the application. This is an example of event-chaining where notification services are used to chain various synchronization events together. This is typically useful when a single thread must process multiple synchronization events.

13.13 Sample System Using a Message Queue for Interthread Communication

We have used counting semaphores for mutual exclusion and for event notification in the two previous sample systems. We have also used an event flags group to synchronize the behavior of two threads. In this sample system, we will use a message queue to

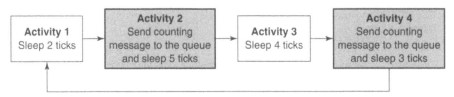

Figure 13.14: Activities of the Speedy_Thread where priority = 5

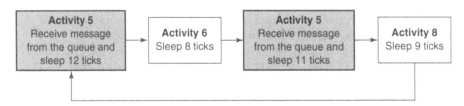

Figure 13.15: Activities of the Slow_Thread where priority = 15

communicate between two threads. We will modify the previous sample system and replace all references to an event flags group with references to a message queue.

In Figure 13.14, when Speedy_Thread enters Activity 2 or Activity 4, it attempts to send one counting message (i.e., 0, 1, 2, 3, ...) to the queue, but if the queue is full, it waits until space becomes available. Speedy_Thread has the same priority and similar activities as in the previous sample system.

In Figure 13.15, when Slow_Thread enters Activity 5 or Activity 7, it attempts to receive one message from the queue, but if the queue is empty, it waits until a message appears. Slow_Thread does not process the value of the message it receives; it simply removes the message from the queue and continues executing. Slow_Thread has the same priority and similar activities as in the previous sample system.

We will design our message queue so that it can store a maximum of 100 messages. In the sample output for this system, the Speedy_Thread completes many more cycles than the Slow_Thread. However, when the queue becomes full, each thread completes the same number of cycles.

We will discuss a series of changes to be applied to the sample system from Chapter 12 so that all references to an event flags group will be replaced with references to a message queue. The complete program list called *13_sample_system.c* is located in the next section of this chapter and on the attached CD.

The first change occurs in the declaration and definitions section of our program, to which we need to add the following *#defines*:

```
#define QUEUE_MSG_SIZE TX_1_ULONG
#define QUEUE_TOTAL_SIZE QUEUE_SIZE*sizeof(ULONG)*QUEUE_MSG_SIZE
```

These *#defines* specify the message size (in ULONGs, not bytes) and the total size of the message queue in bytes. The second *#define* provides some flexibility so that if either the message size or queue capacity (number of messages) were changed, then the total queue size would be calculated accordingly.

We need to replace the declaration of an event flags group with the declaration of a message queue as follows:

```
TX_QUEUE my_queue;
```

We also need to delete the declarations for the event flags group, and specify several new declarations so that we can send and receive our messages, as follows:

```
ULONG send_message[QUEUE_MSG_SIZE]={0x0},
received_message[QUEUE_MSG_SIZE];
```

The next change occurs in the application definitions section of our program, in which we replace the creation of an event flags group with the creation of a message queue, as follows:

```
/* Create the message queue used by both threads. */
tx_queue_create (&my_queue, "my_queue", QUEUE_MSG_SIZE,
                 queue_storage, QUEUE_TOTAL_SIZE);
```

The remaining changes occur in the function definitions section of our program. We need to change all references to an event flags group with references to a message queue. We will show only the changes for the Speedy_Thread and will leave the Slow_Thread changes as an exercise for the reader. Figure 13.16 contains the necessary changes for Activity 2.

Figure 13.17 contains the necessary changes for Activity 4. Most of the modifications involve changing references to an event flags group with references to a message queue.

```
/* Activity 2:  send a message to the queue,
   then sleep 5 timer-ticks.  */

send_message[QUEUE_MSG_SIZE-1]++;

status = tx_queue_send (&my_queue, send_message,
        TX_WAIT_FOREVER);

if (status != TX_SUCCESS)  break;  /* Check status */

tx_thread_sleep(5);
```

Figure 13.16: Changes to Activity 2

```
/* Activity 4:  send a message to the queue,
   then sleep 3 timer-ticks.  */

send_message[QUEUE_MSG_SIZE-1]++;

status = tx_queue_send (&my_queue, send_message,
        TX_WAIT_FOREVER);

if (status != TX_SUCCESS)  break;  /* Check status */

tx_thread_sleep(3);
```

Figure 13.17: Changes to Activity 4

13.14 Listing for 13_sample_system.c

```
001  /* 13_sample_system.c
002
003     Create two threads, one byte pool, and one message queue.
004     The threads communicate with each other via the message
           queue.
005     Arrays are used for the stacks and the queue storage space */
006
007
008  /***************************************************/
009  /*      Declarations, Definitions, and Prototypes     */
010  /***************************************************/
011
012  #include "tx_api.h"
013  #include <stdio.h>
014
```

```
015  #define STACK_SIZE        1024
016  #define QUEUE_SIZE        100
017  #define QUEUE_MSG_SIZE    TX_1_ULONG
018  #define QUEUE_TOTAL_SIZE  QUEUE_SIZE*sizeof(ULONG)*QUEUE_MSG_
                               SIZE
019
020  /* Define thread stacks */
021  CHAR stack_speedy[STACK_SIZE];
022  CHAR stack_slow[STACK_SIZE];
023  CHAR queue_storage[QUEUE_TOTAL_SIZE];
024
025  /* Define the ThreadX object control blocks */
026
027  TX_THREAD Speedy_Thread;
028  TX_THREAD Slow_Thread;
029
030  TX_TIMER stats_timer;
031
032  TX_QUEUE my_queue;
033
034
035  /* Define the counters used in the PROJECT application... */
036
037  ULONG Speedy_Thread_counter=0, total_speedy_time=0;
038  ULONG Slow_Thread_counter=0, total_slow_time=0;
039  ULONG send_message[QUEUE_MSG_SIZE]={0x0},
            received_message[QUEUE_MSG_SIZE];
040
041
042
043  /* Define thread prototypes. */
044
045  void Speedy_Thread_entry(ULONG thread_input);
046  void Slow_Thread_entry(ULONG thread_input);
047  void print_stats(ULONG);
048
049
050  /***************************************************/
051  /*                Main Entry Point                 */
052  /***************************************************/
```

```
053
054   /* Define main entry point. */
055
056   int main()
057   {
058
059     /* Enter the ThreadX kernel. */
060     tx_kernel_enter();
061   }
062
063
064
065   /***************************************************/
066   /*              Application Definitions          */
067   /***************************************************/
068
069
070   /* Define what the initial system looks like. */
071
072   void tx_application_define(void *first_unused_memory)
073   {
074
075     /* Put system definition stuff in here, e.g., thread creates
076        and other assorted create information. */
077
078     /* Create the Speedy_Thread. */
079     tx_thread_create(&Speedy_Thread, "Speedy_Thread",
080                      Speedy_Thread_entry, 0,
081                      stack_speedy, STACK_SIZE, 5, 5,
082                      TX_NO_TIME_SLICE, TX_AUTO_START);
083
084     /* Create the Slow_Thread */
085     tx_thread_create(&Slow_Thread, "Slow_Thread",
086                      Slow_Thread_entry, 1,
087                      stack_slow, STACK_SIZE, 15, 15,
088                      TX_NO_TIME_SLICE, TX_AUTO_START);
089
090
091     /* Create the message queue used by both threads. */
092
```

```
093     tx_queue_create (&my_queue, "my_queue", QUEUE_MSG_SIZE,
094                      queue_storage, QUEUE_TOTAL_SIZE);
095
096
097     /* Create and activate the timer */
098     tx_timer_create (&stats_timer, "stats_timer", print_stats,
099                      0x1234, 500, 500, TX_AUTO_ACTIVATE);
100
101     }
102
103
104     /****************************************************/
105     /*               Function Definitions            */
106     /****************************************************/
107
108
109     /* Entry function definition of the "Speedy_Thread"
110        it has a higher priority than the "Slow_Thread" */
111
112  void Speedy_Thread_entry(ULONG thread_input)
113  {
114
115  UINT status;
116  ULONG start_time, cycle_time=0, current_time=0;
117
118
119     /* This is the higher priority "Speedy_Thread"-it sends
120        messages to the message queue */
121     while(1)
122     {
123
124       /* Get the starting time for this cycle */
125       start_time = tx_time_get();
126
127       /* Activity 1: 2 timer-ticks. */
128       tx_thread_sleep(2);
129
130       /* Activity 2: send a message to the queue, then sleep 5
131                     timer-ticks. */
132       send_message[QUEUE_MSG_SIZE-1]++;
```

```
132
133      status=tx_queue_send (&my_queue, send_message,
                            TX_WAIT_FOREVER);
134
135   if (status !=TX_SUCCESS) break; /* Check status */
136
137   tx_thread_sleep(5);
138
139   /* Activity 3: 4 timer-ticks. */
140   tx_thread_sleep(4);
141
142   /* Activity 4: send a message to the queue, then sleep 3
         timer-ticks */
143   send_message[QUEUE_MSG_SIZE-1]++;
144
145   status=tx_queue_send (&my_queue, send_message,
          TX_WAIT_FOREVER);
146
147   if (status !=TX_SUCCESS) break; /* Check status */
148
149   tx_thread_sleep(3);
150
151
152   /* Increment the thread counter and get timing info */
153   Speedy_Thread_counter++;
154
155   current_time=tx_time_get();
156   cycle_time=current_time - start_time;
157   total_speedy_time=total_speedy_time+cycle_time;
158
159   }
160  }
161
162  /************************************************************
/
163
164  /* Entry function definition of the "Slow_Thread"
165      it has a lower priority than the "Speedy_Thread" */
166
167  void Slow_Thread_entry(ULONG thread_input)
```

```
168   {
169
170   UINT status;
171   ULONG start_time, current_time=0, cycle_time=0;
172
173
174      /* This is the lower priority "Slow_Thread"-it receives
             messages
175         from the message queue */
176      while(1)
177      {
178
179         /* Get the starting time for this cycle */
180         start_time=tx_time_get();
181
182         /* Activity 5 - receive a message from the queue and
                sleep 12 timer-ticks.*/
183         status=tx_queue_receive (&my_queue, received_message,
                                    TX_WAIT_FOREVER);
184
185         if (status !=TX_SUCCESS) break; /* Check status */
186
187         tx_thread_sleep(12);
188
189         /* Activity 6: 8 timer-ticks. */
190         tx_thread_sleep(8);
191
192         /* Activity 7: receive a message from the queue and
                sleep 11 timer-ticks.*/
193
194         /* receive a message from the queue */
195         status=tx_queue_receive (&my_queue, received_message,
                                    TX_WAIT_FOREVER);
196
197         if (status !=TX_SUCCESS) break; /* Check status */
198
199         tx_thread_sleep(11);
200
201         /* Activity 8: 9 timer-ticks. */
202         tx_thread_sleep(9);
```

```
203
204        /* Increment the thread counter and get timing info */
205        Slow_Thread_counter++;
206
207        current_time=tx_time_get();
208        cycle_time=current_time-start_time;
209        total_slow_time=total_slow_time+cycle_time;
210
211    }
212  }
213
214  /******************************************************/
215  /* print statistics at specified times */
216  void print_stats (ULONG invalue)
217  {
218     ULONG current_time, avg_slow_time, avg_speedy_time;
219
220     if ((Speedy_Thread_counter>0) && (Slow_Thread_counter>0))
221     {
222        current_time=tx_time_get();
223        avg_slow_time=total_slow_time/Slow_Thread_counter;
224        avg_speedy_time=total_speedy_time/Speedy_Thread_counter;
225
226        printf("\n**** Threads communicate with a message
                  queue.\n\n");
227        printf(" Current Time:            %lu\n", current_time);
228        printf(" Speedy_Thread counter:   %lu\n",
                  Speedy_Thread_counter);
229        printf(" Speedy_Thread avg time: %lu\n",
                  avg_speedy_time);
230        printf(" Slow_Thread counter:     %lu\n",
                  Slow_Thread_counter);
231        printf(" Slow_Thread avg time:    %lu\n",
                  avg_slow_time);
232        printf(" # messages sent:         %lu\n\n",
233                  send_message[QUEUE_MSG_SIZE-1]);
234     }
235        else printf("Bypassing print_stats function, Current
                     Time: %lu\n",
236                  tx_time_get());
237  }
```

13.15 Message Queue Internals

When the TX_QUEUE data type is used to declare a message queue, a Queue Control Block (QCB) is created, and that Control Block is added to a doubly linked circular list, as illustrated in Figure 13.18.

The pointer named tx_queue_created_ptr points to the first Control Block in the list. See the fields in the QCB for timer attributes, values, and other pointers.

In general, the tx_queue_send and tx_queue_front_send operations copy the contents of a message to a position in the message queue, i.e., to the rear or the front of the queue, respectively. However, if the queue is empty and another thread is suspended because it is waiting for a message, then that message bypasses the queue entirely and goes directly to the destination specified by the other thread. ThreadX uses this shortcut to enhance the overall performance of the system.

13.16 Overview

Message queues provide a powerful tool for interthread communication. Message queues do not support a concept of ownership, nor is there a limit to how many threads can access a queue. Any thread can send a message to a queue and any thread can receive a message from a queue.

If a thread attempts to send a message to a full queue, then its behavior will depend on the specified wait option. These options will cause the thread either to abort the message transmission or to suspend (indefinitely or for a specific number of timer-ticks) until adequate space is available in the queue.

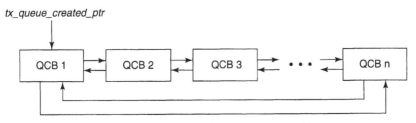

Figure 13.18: Created message queue list

	Thread synchronization	Event notification	Mutual exclusion	Inter-Thread communication
Mutex			Preferred	
Counting Semaphore	OK—better for one event	Preferred	OK	
Event Flags Group	Preferred	OK		
Message Queue	OK	OK		Preferred

Figure 13.19: Recommended uses of public resources

Similarly, if a thread attempts to receive a message from an empty queue, it will behave according to the specified wait option.

Normally, messages on a queue behave in a FIFO manner, i.e., the first messages sent to the rear of the queue are the first to be removed from the front. However, there is a service that permits a message to be sent to the front of the queue, rather than to the rear of the queue.

A message queue is one type of public resource, meaning that it is accessible by any thread. There are four such public resources, and each has features that are useful for certain applications. Figure 13.19 compares the uses of message queues, mutexes, counting semaphores, and event flags groups. As this comparison suggests, the message queue is ideally suited for interthread communication.

13.17 Key Terms and Phrases

flush contents of a queue

front of queue

interthread communication

mailbox

message capacity

message queue

Message Queue Control Block

message queue creation

message queue deletion

message queue FIFO discipline

message size

prioritize a message queue suspension list

Queue Control Block (QCB)

queue storage space

queue suspension

rear of queue

receive message

send message

13.18 Problems

1. Describe how you would implement the producer-consumer system discussed in Chapter 11 so that it would use a message queue rather than a counting semaphore. State all your assumptions.

2. Suppose that you want to synchronize the operation of two threads by using a message queue. Describe what you would have to do in order to make certain that the threads take turns sharing the processor, i.e., thread 1 would access the processor, then thread 2, then thread 1, and so on.

3. Describe how you would determine how many messages are currently stored in a particular message queue.

4. Normally, messages are inserted at the rear of the queue, and are removed from the front of the queue. Describe a scenario in which you should insert messages at the front of the queue instead of the rear of the queue.

5. Suppose that three numbered threads (i.e., 1, 2, 3) use a message queue to communicate with each other. Each message consists of four words (i.e., TX_4_ULONG) in which the first word contains the thread number for which the message is intended, and the other words contain data. Describe how the message queue can be used so that one thread can send a message to any one of the other three threads, and a thread will remove a message from the queue only if it is intended for that thread.

CHAPTER 14

Case Study: Designing a Multithreaded System

14.1 Introduction

The purpose of this chapter is to develop a case study based on an application that could use both the ThreadX RTOS and the ARM processor. The application we will consider is a real-time video/audio/motion (VAM) recording system that could be useful for numerous commercial motorized vehicle fleets around the world.[1]

The VAM system features a small recording device that could be attached to a vehicle's windshield directly behind the rear-view mirror to avoid intrusion into the driver's field of vision. When triggered by an accident or unsafe driving, the VAM system automatically records everything the driver sees and hears in the 12 seconds preceding and the 12 seconds following the event. Events are stored in the unit's digital memory, along with the level of G-forces on the vehicle. In the event of an accident, unsafe driving, warning, or other incident, the VAM system provides an objective, unbiased account of what actually happened.

To complete the system, there should be a driving feedback system that downloads the data from the VAM system unit and provides playback and analysis. This system could also be used to create a database of incidents for all the drivers in the vehicle fleet. We will not consider that system; instead, we will focus on the capture of real-time data for the VAM system unit.

---

[1]The VAM system is a generic design and is not based on any actual implementation.

As noted earlier, most of the VAM unit could be located behind the rear-view mirror, so it would not obscure the vision of the driver. The unit would have to be installed so that the lenses have a clear forward view and rear view. The system includes a readily accessible *emergency button* so the driver can record an unusual, serious, or hazardous incident whenever necessary. The VAM system is constantly recording everything the driver sees, hears, and feels. That is, the VAM system records all visual activities (front and rear of the vehicle), audible sounds, and G-forces.

As an illustration of how the VAM system could be used, consider the following scenario. A driver has been inattentive and has failed to stop at a red traffic light. By the time the driver realizes the error, the vehicle has already entered the intersection and is headed toward an oncoming vehicle. The driver vigorously applies the brakes and swerves to the right. Typical G-forces for this incident are about -0.7 (forward) and $+0.7$(side). Thus, the VAM system detects this incident and records it as an unsafe driving event in the protected memory.

When we download and analyze the data from the VAM system, we should be able to clearly see that the driver ran a red light and endangered passengers and other people on the highway, as well as the vehicle itself. The driver's employer would have been legally liable for the driver's actions if this incident had resulted in a collision.

In this scenario, no collision resulted from this incident. However, this recording would show that this driver was clearly at fault and perhaps needs some refresher training. Figure 14.1 illustrates the G-forces that can be detected, where the front of the vehicle appears at the top of the illustration.

The system stores the 24 seconds of video, audio, and motion recording that surround the time of this incident in protected memory and illuminates a red light that indicates a driving incident has occurred. This light can be turned off only when the special downloading process has been performed; the driver cannot turn it off.

We will design the VAM system with the ThreadX RTOS and the ARM processor. For simplicity, we will omit certain details that are not important to the development of this system, such as file-handling details.[2]

---

[2]We could use a software companion to ThreadX that could handle those file operations. That software product is *FileX*, but discussing it is beyond the scope of this book.

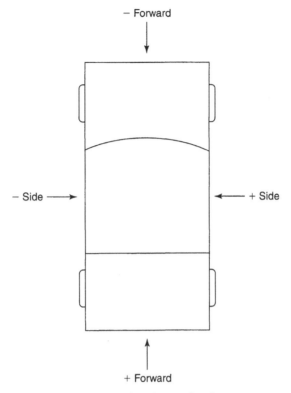

Figure 14.1: Directions of G-forces

14.2 Statement of Problem

The VAM system is based on a set of sensors that measure G-forces experienced by a driver in a motorized vehicle. The system uses two sets of measurements. One set indicates forward or backward motion of the vehicle. Negative forward values indicate deceleration, or G-forces pushing against the driver's front side, while positive forward values indicate acceleration, or G-forces pushing against the driver's back. The other set of measurements indicates sideways motion of the vehicle. Negative side values indicate acceleration to the right, or G-forces pushing against the driver's left side, while positive side values indicate acceleration to the left, or G-forces pushing against the driver's right side. For example, if a vehicle makes a hard left turn, then the sensors produce a positive side value.

Event	Priority
Crash	1
Unsafe driving	2
Warning	3
Manually triggered	4

Figure 14.2: Events and corresponding priorities

The VAM system detects and reports four categories of events. We assign each category a priority,[3] indicating the importance of the event. Figure 14.2 lists the event categories and their corresponding priorities.

Event priorities serve two primary purposes. First, a priority indicates the severity of an event. Second, an event priority determines whether the current event can overwrite a previously stored event in the protected memory. For example, assume that the protected memory is full and the driver hits the *emergency button*, thereby creating a manually triggered event. The only way that this event can be saved is if a previous manually triggered event has already been stored. Thus, a new event can overwrite a stored event of the same or lower priority, but it cannot overwrite a stored event with a higher priority.

If the G-force sensors detect an accident, unsafe driving, or warning, the VAM system generates an interrupt so that ThreadX can take appropriate action and archive that event. Figure 14.3 contains a graphical representation of the G-forces in this system, in which the event labeled by the letter "W" is a warning event.

The driver may hit the *emergency button* at any time to generate an interrupt, signifying a manually triggered event.

Figure 14.4 contains the actual G-force values that are used to detect and report these events. We assume symmetry in how we classify forward and side G-forces, but we could easily modify that assumption without affecting our design.

---

[3]This event priority is not the same as a ThreadX thread or interrupt priority. We use event priorities to classify the relative importance of the events; we do not use them to affect the time when the events are processed.

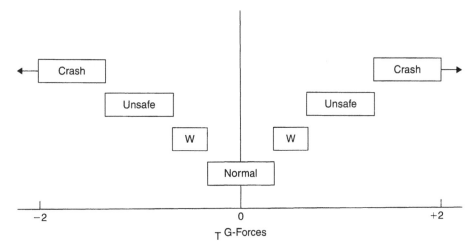

Figure 14.3: Graphical classification of events by G-forces

Event	Forward G-Force	Side G-Force
Crash	+1.6 ≤ Forward	+1.6 ≤ Side
Unsafe driving	+1.6 > Forward >= +0.7	+1.6 > Side >= +0.7
Warning	+0.7 > Forward >= +0.4	+0.7 > Side >= +0.4
Normal driving	+0.4 > Forward > −0.4	+0.4 > Side > −0.4
Warning	−0.4 >= Forward > −0.7	−0.4 >= Side > −0.7
Unsafe driving	−0.7 >= Forward > −1.6	−0.7 >= Side > −1.6
Crash	−1.6 >= Forward	−1.6 >= Side

Figure 14.4: G-Forces and event classifications

To add some perspective about G-forces, consider a vehicle accelerating from zero to 60 miles per hour (0 to 96 kilometers/hour) in six seconds. This produces a G-force of about 0.4—not enough to trigger an unsafe incident report, but enough to trigger a warning event. However, if a driver is applying "hard braking" to a vehicle, it could produce a

G-force of about 0.8, which would trigger an unsafe driving event. If a vehicle crashes into a solid wall while traveling at 62 mph (100 km/hr), this produces a G-force of almost 100!

The VAM system uses two nonvolatile memory systems: a temporary memory system and a protected memory system. The protected memory system stores only detected or manually triggered incidents, while the other system is a temporary memory that records video, audio, and G-forces. It's not necessary to retain ordinary driving activities, so the temporary memory system is overwritten after some period of time, depending on the size of the temporary memory. As noted previously, the protected memory system stores all crash events, unsafe driving events, warnings, and manually triggered events, plus associated audio and video, as long as memory space is available. The protected memory system could be available in several different sizes, and our design will be able to accommodate those different memory sizes.

14.3 Analysis of the Problem

Figure 14.5 illustrates the temporary memory system used for continuous recording. This is actually a circular list where the first position logically follows the last position in the list.

This system provides temporary storage that is overwritten repeatedly. Its main purpose is to provide data storage for an event that needs to be saved in the protected memory. When an event occurs, the 12 seconds preceding the event have already been stored in the temporary memory. After the 12 seconds of data following the event have been stored, the system stores this 24 seconds of data in protected memory.

The actual size of the temporary memory can be configured to the needs of the user. Figure 14.6 illustrates the protected memory that is used to store the automatically detected or manually triggered events.

The size of the protected memory can also be configured according to the needs of the user. We arbitrarily assume that this memory can store 16 events, although we can

Beginning of memory   End of memory

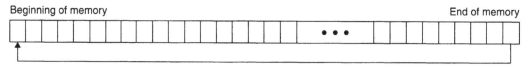

Figure 14.5: Temporary memory (Circular list)

change this value without affecting our design. In addition to the actual data for the event, the protected memory must store the priority and time of the event's occurrence. This information is essential in the case where the protected memory becomes full and another event is detected or manually triggered. An event can never overwrite a higher-priority event, but it can overwrite an event with the same or lower priority. Figure 14.7 summarizes the event overwrite rules.

As stated previously, when the G-force sensors detect a crash, the system generates an interrupt with event priority 1. Unsafe driving events are logged as event priority 2,

Figure 14.6: Protected memory

Priority	Overwrite Rule When Protected Memory is Full
1	Overwrite the oldest event of Priority 4. If no Priority 4 event exists, overwrite the oldest event of Priority 3. If no Priority 3 event exists, overwrite the oldest event of Priority 2. If no Priority 2 event exists, overwrite the oldest event of Priority 1.
2	Overwrite the oldest event of Priority 4. If no Priority 4 event exists, overwrite the oldest event of Priority 3. If no Priority 3 event exists, overwrite the oldest event of Priority 2. If no Priority 2 event exists, do not save the new event.
3	Overwrite the oldest event of Priority 4. If no Priority 4 event exists, overwrite the oldest event of Priority 3. If no Priority 3 event exists, do not save the new event.
4	Overwrite the oldest event of Priority 4. If no Priority 4 event exists, do not save the new event.

Figure 14.7: Event overwrite rules

warnings as event priority 3, and manual events (pushing the *emergency button*) as event priority 4. Our objective is to respond to these events (which will appear to the system as interrupts), as well as to handle initialization and to process routine data.

14.4 Design of the System

Our design will be simplified and will concentrate on control issues. In this section, we will consider thread design issues, and public resources design. We will assume that other processes will handle[4] the actual storing and copying of data.

14.4.1 Thread Design

We need a thread to perform various initialization duties such as setting pointers and variables, opening the files for the temporary memory and the protected memory, and establishing communication paths and buffers. We will assign the name *initializer* to this thread. Figure 14.8 depicts this thread.

We need a thread to coordinate the capture and storage of data from the VAM system unit to the temporary memory. This thread manages the transfer of audio, video, and G-force data from the VAM system unit to an internal buffer, and then to the temporary memory. We will assign the name *data_capture* to this thread. Figure 14.9 illustrates this thread's operation.

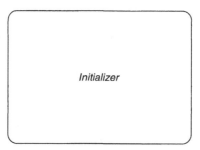

Initializer

Figure 14.8: Initialization of the VAM system

---

[4]This is a good use of a memory byte pool because space is allocated from the pool only once, thus eliminating fragmentation problems. Furthermore, the memory allocations are of different sizes.

We need four ISRs to handle the detected and triggered events. We need one message queue called *event_notice* to store information about an event until the copying process begins. We also need one thread to process the copying of data from the temporary memory to the protected memory. We will assign the name *event_recorder* to this thread. This thread is activated 12 seconds after an event has been detected. At this time, the temporary memory contains the 12 seconds of data preceding the event and 12 seconds of data following the event. The *event_recorder* thread then takes identifying information from the *event_notice* queue and then copies the audio, video, G-force, and timing data for this event from the temporary memory to the protected memory, including the event priority. If protected memory is full, the event *recorder thread* employs the overwrite rules shown in Figure 14.7.

We will use four timers to simulate the arrival of external interrupts for the detectable and triggered events. To simulate such an interrupt, the timer in question saves the thread context, invokes the corresponding ISR, and then restores the thread context. We will assign the names *crash_interrupt, unsafe_interrupt, warning_interrupt,* and *manual_interrupt* to these four timers. Figure 14.10 presents an overview of the interaction between the timers that simulate interrupts, their associated ISRs, the scheduling timers, and the event recorder thread that actually performs the copying.

14.4.2 Public Resources Design

We will use one mutex to provide protection for the protected memory while copying data from the temporary memory. To begin copying data, the *event_recorder* thread must obtain ownership of this mutex. We will assign the name *memory_mutex* to this mutex.

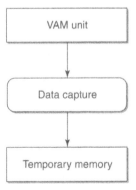

Figure 14.9: Capturing data from the VAM unit

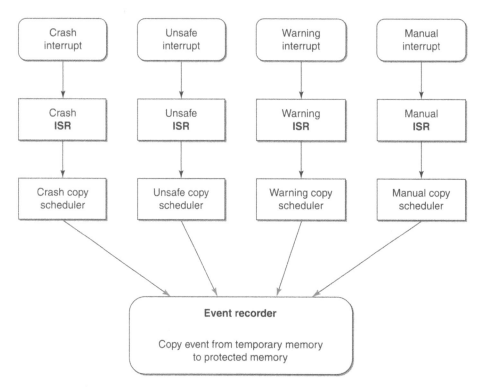

Figure 14.10: Overview of event interrupts and data recording

We need four application timers to schedule the copying of data from temporary memory to protected memory. When an event is detected, the corresponding application timer is activated and expires precisely 12 seconds later. At this point in time, the *event_recorder* thread begins copying data. We will assign the names *crash_timer, unsafe_timer, warning_timer*, and *manual_timer* to these four application timers. Figure 14.11 illustrates how these application timers generate interrupts and schedule the processing of events.

As illustrated by Figure 14.11, an event interrupt is simulated by one of the four timers created for this purpose. One of four corresponding ISRs is invoked and it sends event information to the message queue, activates one of four corresponding scheduling timers, and sets it to expire in 12 seconds. When that timer expires, it *resumes* the *event_recorder* thread, which should be in a suspended state unless it is in the process of copying information from temporary memory to protected memory.

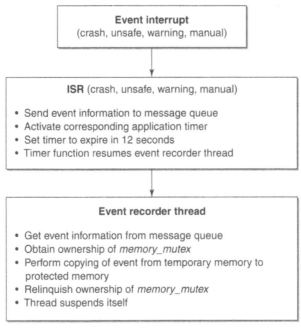

Figure 14.11: Event processing

When the *event_recorder* thread resumes execution, it takes one message from the queue. This message contains two pieces of information about an event: the frame index and the event priority. The frame index is a location in temporary memory that specifies where information about the event begins. The *event_recorder* thread then obtains the *memory_mutex* and if successful, proceeds to copy information from temporary memory to protected memory. When the thread completes copying, it releases the mutex and suspends itself.

Figure 14.12 contains a summary of the public resources needed for this case study.

Figure 14.13 contains a summary of the definitions, arrays, and variables used. We will simulate the temporary memory file system and the protected memory file system with arrays.

The event counters in Figure 14.13 are used when printing periodic statistics about the state of the system. We also need a collection of functions to perform the actions of the

Public resource	Type	Description
initializer	Thread	Perform initialization operations
data_capture	Thread	Perform routine capture of data to temporary memory
event_recorder	Thread	Copy event information from temporary memory to protected memory
event_notice	Queue	Message queue containing key information about detected or triggered events
memory_mutex	Mutex	Mutex to guard protected memory during copying
my_byte_pool	byte_pool	Provide memory for thread stacks and message queue
crash_interrupt	Timer	Generate a simulated crash event interrupt at periodic intervals
unsafe_interrupt	Timer	Generate a simulated unsafe event interrupt at periodic intervals
warning_interrupt	Timer	Generate a simulated warning event interrupt at periodic intervals
manual_interrupt	Timer	Generate a simulated manually triggered event interrupt at periodic intervals
crash_copy_scheduler	Timer	Schedule the time at which copying of a crash event should begin
unsafe_copy_scheduler	Timer	Schedule the time at which copying of an unsafe event should begin
warning_copy_scheduler	Timer	Schedule the time at which copying of a warning event should begin
manual_copy_scheduler	Timer	Schedule the time at which copying of a manually triggered event should begin
stats_timer	Timer	Print system statistics at periodic intervals

Figure 14.12: Summary of public resources used for the VAM system

timers and the threads. Figure 14.14 summarizes all thread entry functions and timer expiration functions used in this system.

We will develop the complete program for our system in the next section. We will develop the individual components of our system and then present a complete listing of the system in a later section.

Name	Type	Description
STACK_SIZE	Define	Represents the size of each thread stack
BYTE_POOL_SIZE	Define	Represents the size of the memory byte pool
MAX_EVENTS	Define	Represents the maximum number of events that can be stored in protected memory
MAX_TEMP_MEMORY	Define	Represents the maximum number of data frames that can be stored in temporary memory
temp_memory	ULONG	Array representing temporary memory
protected_memory	ULONG	Array representing protected memory
frame_data	ULONG	Array representing information about an event
frame_index	ULONG	Location in temporary memory where an event occurred
event_count	ULONG	Number of entries in protected memory
num_crashes	ULONG	Counter for the number of crash events
num_unsafe	ULONG	Counter for the number of unsafe events
num_warning	ULONG	Counter for the number of warning events
num_manual	ULONG	Counter for the number of manual events

Figure 14.13: Definitions, arrays, and counters used in the VAM system

14.5 Implementation

Our implementation will be simplified because we are primarily interested in developing a control structure for this system. Thus, we will omit all file-handling details, represent files as arrays, and simulate capture of data once per second. (An actual implemented system would capture data about 20 to 40 times per second.) For convenience, we will represent each clock timer-tick as one second.

For this system, we will display information on the screen to show when events are generated and how they are processed. We will also display summary information on a

Function	Description
initializer_process	Invoked by thread initializer; perform initialization operations
data_capture_process	Invoked by thread data_capture; perform routine capture of data to temporary memory
event_recorder_process	Invoked by thread event_recorder; copy event data from temporary memory to protected memory
crash_ISR	Invoked by timer crash_interrupt when crash event detected; initiates crash event processing
unsafe_ISR	Invoked by timer unsafe_interrupt when unsafe event detected; initiates unsafe event processing
warning_ISR	Invoked by timer warning_interrupt when warning event detected; initiates warning event processing
manual_ISR	Invoked by timer manual_interrupt when manual event triggered; initiates manual event processing
crash_copy_activate	Invoked by timer crash_copy_scheduler 12 seconds after event occurrence; schedules event copying
unsafe_copy_activate	Invoked by timer unsafe_copy_scheduler 12 seconds after event occurrence; schedules event copying
warning_copy_activate	Invoked by timer warning_copy_scheduler 12 seconds after event occurrence; schedules event copying
manual_copy_activate	Invoked by timer manual_copy_scheduler 12 seconds after event occurrence; schedules event copying
print_stats	Invoked by timer print_stats; prints system statistics at periodic intervals

Figure 14.14: Entry and expiration functions used in the VAM system

periodic basis. Figure 14.15 contains sample diagnostic output for our system that we could use during development.

We will arbitrarily schedule a summary of system statistics every 1,000 timer-ticks. Note that in Figure 14.15, warning events occur at times 410 and 820. Recall that a warning event has priority 3. Also, an unsafe event occurs at time 760 and a manually triggered event occurs at time 888. When the system summary is displayed at time 1,000, we note that the four detected events have been stored in protected memory. The repeated

```
VAM System - Trace of Event Activities Begins...

*Event**    Time:    410    Count:  0    Pri: 3
*Event**    Time:    760    Count:  1    Pri: 2
*Event**    Time:    820    Count:  2    Pri: 3
*Event**    Time:    888    Count:  3    Pri: 4


*** VAM System Periodic Event Summary
     Current Time:                  1000
        Number of Crashes:          0
        Number of Unsafe Events:    1
        Number of Warnings:         2
        Number of Manual Events:    1

*** Portion of Protected Memory Contents
   Time   Pri  Data
    410    3   4660   4660   4660   4660   4660   4660   (etc.)
    760    2   4660   4660   4660   4660   4660   4660   (etc.)
    820    3   4660   4660   4660   4660   4660   4660   (etc.)
    888    4   4660   4660   4660   4660   4660   4660   (etc.)
```

Figure 14.15: Sample output produced by VAM system

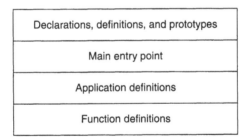

Figure 14.16: Basic structure for the VAM system

data values of 4660 (or 0×1234) are used only to suggest the capture of data from the VAM unit.

We will use the same basic structure for the VAM system as we used for the sample programs in previous chapters. Figure 14.16 contains an overview of this structure.

We will begin by creating the declarations, definitions, and prototypes needed for our system. Figure 14.17 contains the first part of this section. The values we have chosen for the stack size, memory byte pool size, the size of the protected memory, and the size of

```
#include    "tx_api.h"
#include    <stdio.h>

#define     STACK_SIZE          1024
#define     BYTE_POOL_SIZE      9120
#define     MAX_EVENTS            16
#define     MAX_TEMP_MEMORY      200

/* Define the ThreadX object control blocks */

TX_THREAD       initializer;
TX_THREAD       data_capture;
TX_THREAD       event_recorder;

TX_QUEUE        event_notice;

TX_MUTEX        memory_mutex;
TX_BYTE_POOL    my_byte_pool;

TX_TIMER        crash_interrupt;
TX_TIMER        unsafe_interrupt;
TX_TIMER        warning_interrupt;
TX_TIMER        manual_interrupt;

TX_TIMER        crash_copy_scheduler;
TX_TIMER        unsafe_copy_scheduler;
TX_TIMER        warning_copy_scheduler;
TX_TIMER        manual_copy_scheduler;
TX_TIMER        stats_timer;
```

Figure 14.17: Declarations and definitions—Part 1

the temporary memory can be modified if desired. Many of the entries in Figure 14.17 are definitions of public resources, such as threads, timers, the message queue, the mutex, and the memory byte pool.[5]

Figure 14.18 contains the second part of the program section, devoted to declarations, definitions, and prototypes. We declare the counters, variables, and arrays for our system, as well as the prototypes for our thread entry functions, the timer expiration functions, and the function to display periodic system statistics.

---

[5]This is a good use of a memory byte pool because space is allocated from the pool only once, thus eliminating fragmentation problems. Furthermore, the memory allocations are of different sizes.

```
/* Define the counters and variables used in the VAM system  */

ULONG   num_crashes=0, num_unsafe=0, num_warning=0, num_manual=0;
ULONG   frame_index, event_count, frame_data[2];

/* Define the arrays used to represent temporary memory        */
/* and protected memory. temp_memory contains pair of data     */
/* in the form time-data and protected_memory contains rows    */
/* of 26 elements in the form time-priority-data-data-data... */
/* The working index to temp_memory is frame_index and the     */
/* working index to protected_memory is event_count.           */

ULONG   temp_memory[MAX_TEMP_MEMORY][2],
        protected_memory[MAX_EVENTS][26];

/* Define thread and function prototypes.  */

void   initializer_process(ULONG);
void   data_capture_process(ULONG);
void   event_recorder_process(ULONG);
void   crash_ISR(ULONG);
void   unsafe_ISR(ULONG);
void   warning_ISR(ULONG);
void   manual_ISR(ULONG);
void   crash_copy_activate(ULONG);
void   unsafe_copy_activate(ULONG);
void   warning_copy_activate(ULONG);
void   manual_copy_activate(ULONG);
void   print_stats(ULONG);#include  "tx_api.h"
#include  <stdio.h>
```

Figure 14.18: Declarations, definitions, and prototypes—Part 2

The two arrays declared in Figure 14.18 represent file systems that would be used in an actual implementation. The array named *temp_array* represents the temporary memory used by the system. As noted earlier, the primary purpose of temporary memory is to provide an ongoing repository of video, audio, and motion data. When an event is detected or manually triggered, the 12 seconds of data before and after that event are copied to protected memory. Temporary memory is overwritten repeatedly because it is limited in size. Figure 14.19 illustrates the organization of temporary memory.

The array named *protected_memory* represents the protected memory in our system. Each entry in this array represents an event that has occurred. Figure 14.20 illustrates

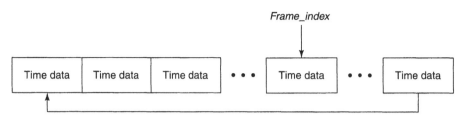

Figure 14.19: Organization of temporary memory

Event time	Event priority	24 second of data for event
Event time	Event priority	24 second of data for event
• • •		
Event time	Event priority	24 second of data for event

Figure 14.20: Organization of protected memory

the organization of protected memory. Each stored event contains the time the event occurred, the event priority, and the captured data for the event.

The size of the protected memory is relatively small and is specified as MAX_EVENT in Figure 14.17. Having a small size is reasonable because the number of vehicle events should also be relatively small.

The main entry point is the same as that used in all the sample systems in the preceding chapters. This is the entry into the ThreadX kernel. Note that the call to tx_kernel_enter does not return, so do not place any processing after it. Figure 14.21 contains the main entry point for the VAM system.

The next portion of our system is the application definitions section, which we will divide into two parts. In the first part, we will define the public resources needed for our system. This includes the memory byte pool, the three threads, the nine timers, and the message queue. Figure 14.22 contains the first part of the applications definitions section.

```
/*****************************************************/
/*                  Main Entry Point                 */
/*****************************************************/

/* Define main entry point.  */

int main()
{

    /* Enter the ThreadX kernel.  */
    tx_kernel_enter();
}
```

Figure 14.21: The main entry point

This part of our system consists of the application definition function called tx_
application_define. This function defines all the application resources in the system.
This function has a single input parameter, which is the first available RAM address. This
is typically used as a starting point for run-time memory allocations of thread stacks,
queues, and memory pools.

The first declaration in Figure 14.22 is the following:

CHAR *byte_pointer;

This pointer is used when allocating memory from the byte pool for the threads and for
the message queue. We then create the memory byte pool, as follows:

tx_byte_pool_create(&my_byte_pool, "my_byte_pool",
                    first_unused_memory, BYTE_POOL_SIZE);

We need to allocate stack space from the byte pool and create the *initializer* thread, as
follows:

tx_byte_allocate(&my_byte_pool, (VOID **) &byte_pointer,
                STACK_SIZE, TX_NO_WAIT);
tx_thread_create(&initializer, "initializer",
                initializer_process, 0, byte_pointer, STACK_SIZE,
                11, 11, TX_NO_TIME_SLICE, TX_AUTO_START);

We assign the initializer thread the highest priority of the three threads in our system.
This thread needs to perform its operations immediately, and then it terminates. The other
two threads continue operating for the life of the system. We create the *data_capture*

```
/**********************************************************/
/*                Application Definitions                 */
/**********************************************************/

/* Define what the initial system looks like.  */

void  tx_application_define(void *first_unused_memory)
{

CHAR    *byte_pointer;

    /* Put system definition stuff in here, e.g., thread creates
       and other assorted create information.  */

    /* Create a memory byte pool from which to allocate
       the thread stacks and the message queue.  */
    tx_byte_pool_create(&my_byte_pool, "my_byte_pool",
       first_unused_memory, BYTE_POOL_SIZE);

    /* Allocate the stack for the initializer thread.  */
    tx_byte_allocate(&my_byte_pool, (VOID **) &byte_pointer,
       STACK_SIZE, TX_NO_WAIT);

    /* Create the initializer thread.  */
    tx_thread_create(&initializer, "initializer",
                    initializer_process, 0,
                    byte_pointer, STACK_SIZE, 11, 11,
                    TX_NO_TIME_SLICE, TX_AUTO_START);

    /* Allocate the stack for the data_capture thread.  */
    tx_byte_allocate(&my_byte_pool, (VOID **) &byte_pointer,
       STACK_SIZE, TX_NO_WAIT);

    /* Create the data_capture thread.  */
    tx_thread_create(&data_capture, "data_capture",
                    data_capture_process, 0,
                    byte_pointer, STACK_SIZE, 15, 15,
                    TX_NO_TIME_SLICE, TX_AUTO_START);

    /* Allocate the stack for the event_recorder thread.  */
    tx_byte_allocate(&my_byte_pool, (VOID **) &byte_pointer,
       STACK_SIZE, TX_NO_WAIT);

    /* Create the event_recorder thread.  */
    tx_thread_create(&event_recorder, "event_recorder",
                    event_recorder_process, 0,
                    byte_pointer, STACK_SIZE, 12, 12,
                    TX_NO_TIME_SLICE, TX_DONT_START);
```

Figure 14.22: Application definitions—Part 1

thread and the *event_recorder* thread in a similar fashion. However, we assign the *event_ recorder* thread a higher priority than the *data_capture* thread because it is essential that the event recording operation be performed in a timely manner.[6] Note that we use neither the time-slicing option nor the preemption-threshold option for any of these threads. However, the *event_recorder* thread is given the TX_DON'T_START option, but the other two threads are given the TX_AUTO_START option.

The second part of the applications definitions section appears in Figure 14.23; this part contains definitions of our timers and the message queue.

We create nine timers and one message queue in Figure 14.23. There are four timers dedicated to simulating interrupts, four timers to schedule event recording, and one timer to display system statistics. We first create the *crash_interrupt* timer, which simulates crash events. We arbitrarily simulate crash events every 1,444 timer-ticks—recall that we simulate one second with one timer-tick. Following is the definition of the *crash_ interrupt* timer.

```
tx_timer_create (&crash_interrupt, "crash_interrupt", crash_ISR,
                0×1234, 1444, 1444, TX_AUTO_ACTIVATE);
```

The expiration function associated with the *crash_interrupt* timer is *crash_ISR*, which we will define in the next section of our system. This function is activated every 1,444 timer-ticks. We do not use a parameter in the expiration function, so we will specify an arbitrary argument value when we create the timer—in this case, the value 0×1234. The *unsafe_interrupt* timer, the *warning_interrupt* timer, and the *manual_interrupt* timer are created in a similar manner. However, we specify different expiration values for each of these timers. Note that we give each of these timers the TX_AUTO_ACTIVATE option.

We create four timers to schedule the copying of an event from temporary memory to protected memory. We first create the *crash_copy_scheduler* timer as follows:

```
tx_timer_create (&crash_copy_scheduler, "crash_copy_scheduler",
                crash_copy_activate, 0×1234, 12, 12,
                TX_NO_ACTIVATE);
```

---

[6]The actual values of the priorities are not important, provided that the *initializer* thread has the highest priority, and the *event_recorder* thread has a higher priority than the *data_capture* thread.

```
/* Create and activate the 4 timers to simulate interrupts */
tx_timer_create (&crash_interrupt, "crash_interrupt", crash_ISR,
                0x1234, 1444, 1444, TX_AUTO_ACTIVATE);
tx_timer_create (&unsafe_interrupt, "unsafe_interrupt", unsafe_ISR,
                0x1234, 760, 760, TX_AUTO_ACTIVATE);
tx_timer_create (&warning_interrupt, "warning_interrupt", warning_ISR,
                0x1234, 410, 410, TX_AUTO_ACTIVATE);
tx_timer_create (&manual_interrupt, "manual_interrupt", manual_ISR,
                0x1234, 888, 888, TX_AUTO_ACTIVATE);

/* Create and activate the 4 timers to initiate data copying */
tx_timer_create (&crash_copy_scheduler, "crash_copy_scheduler",
                crash_copy_activate, 0x1234, 12, 12, TX_NO_ACTIVATE);
tx_timer_create (&unsafe_copy_scheduler, "unsafe_copy_scheduler",
                unsafe_copy_activate, 0x1234, 12, 12, TX_NO_ACTIVATE);
tx_timer_create (&warning_copy_scheduler, "warning_copy_scheduler",
                warning_copy_activate, 0x1234, 12, 12, TX_NO_ACTIVATE);
tx_timer_create (&manual_copy_scheduler, "manual_copy_scheduler",
                manual_copy_activate, 0x1234, 12, 12, TX_NO_ACTIVATE);

/* Create and activate the timer to print statistics periodically */
tx_timer_create (&stats_timer, "stats_timer", print_stats,
                0x1234, 1000, 1000, TX_AUTO_ACTIVATE);

/* Create the message queue that holds the indexes for all events.   */
/* The frame_index is a position marker for the temp_memory array.    */
/* When an event occurs, the event ISR sends the current frame_index  */
/* and event priority to the queue to store crash event information.  */
/* Allocate memory space for the queue, then create the queue.        */

tx_byte_allocate(&my_byte_pool, (VOID **) &byte_pointer,
                MAX_EVENTS*2*sizeof(ULONG), TX_NO_WAIT);
tx_queue_create (&event_notice, "event_notice", TX_2_ULONG,
                byte_pointer, MAX_EVENTS*2*sizeof(ULONG));
}
```

Figure 14.23: Application definitions—Part 2

When an event is detected (or generated, in our system), the associated ISR activates the corresponding scheduling timer—note that the TX_NO_ACTIVATE option is specified for each of the scheduling timers. Each of the scheduling timers has an expiration value of 12 timer-ticks, so it will expire exactly 12 timer-ticks after an event has been detected. As before, we pass a dummy argument to the expiration function. The *unsafe_copy_ scheduler* timer, the *warning_copy_scheduler* timer, and the *manual_copy_scheduler* timer are each created in a similar fashion.

The timer to print system statistics on a periodic basis is created as follows:

```
tx_timer_create (&stats_timer, "stats_timer", print_stats,
                 0X1234, 1000, 1000, TX_AUTO_ACTIVATE);
```

We create this timer with the TX_AUTO_ACTIVATE option, so it expires every 1,000 timer-ticks. When this timer does expire, it invokes the expiration function *print_stats*, which displays the summary statistics.

The last definition in Figure 14.23 creates the message queue *event_notice*. We must first allocate space from the memory byte pool for the message queue, and then create the queue. Following is the creation of this queue:

```
tx_byte_allocate(&my_byte_pool, (VOID **) &byte_pointer,
                 MAX_EVENTS*2*sizeof(ULONG), TX_NO_WAIT);
tx_queue_create (&event_notice, "event_notice", TX_2_ULONG,
                 byte_pointer, MAX_EVENTS*2*sizeof(ULONG));
```

The expression

```
MAX_EVENTS*2*sizeof(ULONG)
```

computes the amount of space needed by the message queue, based on the maximum number of events in protected memory, the size of each message, and the size of the ULONG data type. We use this same expression to specify the amount of space for the queue when it is created. The tx_byte_allocate service creates a block of bytes specified by this expression; the block pointer that receives the address of this block is pointed to by *&byte_pointer*. We specify the TX_NO_WAIT option as the wait option.[7]

The TX_2_ULONG option specifies the size of each message in the queue. The location of memory space for the queues is pointed to by *byte_pointer*. Figure 14.24 illustrates the contents of the *event_notice* message queue.

The frame index contains the location in temporary memory where the event began. The event priority indicates the type of event that occurred.

The last portion of our system is the function definitions section, which we will divide into five parts. In the first part, we will define the *initializer* entry function, which

---

[7]The TX_NO_WAIT option is the only valid wait option when the tx_byte_allocate service is called from initialization.

Figure 14.24: Message queue event_notice

```
/* Entry function definition of the initializer thread */

void  initializer_process(ULONG thread_input)
{
/* Perform initialization tasks                          */
/* Because we are using arrays to represent files, there is  */
/* very little initialization to perform. We initialize two  */
/* global variables that represent starting array indexes.   */
    printf("VAM System - Trace of Event Activities Begins...\n\n");
    frame_index=0;
    event_count=0;
}
```

Figure 14.25: Function definitions Part 1—initializer entry function

performs basic initialization operations. The second part contains the definition of the *data_capture_process* entry function, which simulates data capture. The third part contains definitions of the *crash_ISR* expiration function and the *crash_copy_activate* expiration function. The fourth part contains the definition of the *event_recorder_process* entry function, which copies an event from temporary memory to protected memory. The fifth part contains the *print_stats* expiration function, which displays system statistics at periodic intervals.

Figure 14.25 contains the definition for the *initializer* entry function. This is the first function that is executed. This function merely initializes the *frame_index* and *event_count* global variables. These variables point to the current position of temporary memory and protected memory, respectively. If an actual file system were used, the *initializer* function would have considerably more work to do.

Figure 14.26 contains the definition for the *data_capture_process* entry function. This function runs constantly and simulates the capture of data from the VAM unit every

```
/* Entry function definition of the data_capture thread */

void    data_capture_process(ULONG thread_input)
{
/* Perform data capture from the VAM system to temporary memory  */
/* This function simulates the data capture operation by writing */
/* to an array, which represents a temporary memory file. For    */
/* simplicity, we will write to the array once every timer-tick  */
    while (1)  {
        temp_memory[frame_index][0] = tx_time_get();
        temp_memory[frame_index][1] = 0x1234;
        frame_index = (frame_index + 1) % MAX_TEMP_MEMORY;
        tx_thread_sleep(1);
    }
}
```

Figure 14.26: Function definitions Part 2—data_capture_process entry function

timer-tick. We will use the value 0×1234 to suggest data that is being captured and we will get the current value of the system clock. We will then store those two values in the *temp_memory* array, which represents the temporary memory file. The value of the *frame_index* is advanced by one during each timer-tick, and it wraps around to the beginning when it encounters the last position of the array.

Figure 14.27 contains definitions for the *crash_ISR* expiration function and the *crash_copy_activate* expiration function. These two functions are placed together because they are closely related to each other. The *crash_ISR* function sends the *frame_index* value and the priority (i.e., 1 for a crash) to the *event_notice* message queue. It then activates the *crash_copy_scheduler* timer, which expires in 12 seconds and then invokes the *crash_copy_activate* expiration function. The *crash_copy_activate* function resumes the *event_recorder* thread and then deactivates the *crash_copy_scheduler* timer. We do not show the corresponding expiration functions for the unsafe, warning, and manual events because they are quite similar. One difference is that different priority values are associated with these events.

Figure 14.28 contains the definition for the *event_recorder_process* entry function, which is part of the *event_recorder thread*. When this thread is resumed by one of the scheduler timers, this function copies a crash event, an unsafe event, a warning event, or a

```
/*************************************************************/
/********** crash event detection and processing ***********/
/*************************************************************/

/* Timer function definition for simulated crash interrupt    */
/* This is a simulated ISR -- an actual ISR would probably begin */
/* with a context save and would end with a context restore.   */

void    crash_ISR (ULONG timer_input)
{
   ULONG frame_data[2];
   frame_data[0] = frame_index;
   frame_data[1] = 1;
   num_crashes++;
/* Activates timer to expire in 12 seconds - end of event */
/* Put frame_index and priority on queue for crash events */
   tx_queue_send (&event_notice, frame_data, TX_NO_WAIT);
   tx_timer_activate (&crash_copy_scheduler);
}

/*************************************************************/
/* Timer function definition for timer crash_copy_scheduler,  */
/* which resumes thread that performs recording of crash data */

void    crash_copy_activate (ULONG timer_input)
{
    /* resume recorder thread to initiate data recording */
    tx_thread_resume (&event_recorder);
    tx_timer_deactivate (&crash_copy_scheduler);
}
```

Figure 14.27: Function definitions Part 3—crash_ISR and crash_copy_scheduler expiration functions.

manual event from temporary memory to protected memory. Following are the operations performed by this function:

1. Get the first message on the queue.

2. Extract the frame index and priority for the event from the message.

3. Get the frame in temporary memory (containing the time and captured data) pointed to by the frame index.

4. Compute the copy starting point, which is 12 seconds before the current frame.

5. If protected memory is not full, do the following:

a. Get the mutex.

b. Store the event time and priority in the next available position in protected memory.

c. Store the 24 seconds of frame data from temporary memory to protected memory.

d. Increment the event counter.

e. Release the mutex.

```
/***************************************************************/
/**** Entry function definition of thread event_recorder ****/
/***************************************************************/

void     event_recorder_process(ULONG thread_input)
{
    ULONG frame, event_priority, event_time, index, frame_data[2];
    while (1)
    {
/* Copy an event from temporary memory to protected memory.   */
/* Get frame_index from event message queue and copy 24 frames */
/* from temp_memory to protected_memory.                       */
        tx_queue_receive (&event_notice, frame_data, TX_NO_WAIT);
        /* Store event time and event priority in protected memory */
        frame           = frame_data[0];
        event_priority  = frame_data[1];
        event_time      = temp_memory[frame][0];
        printf("**Event**    Time: %5lu    Count: %2lu    Pri: %lu",
                event_time, event_count, event_priority);
        if (event_count < MAX_EVENTS)
        {
            tx_mutex_get (&memory_mutex, TX_WAIT_FOREVER);
            protected_memory[event_count][0] = event_time;
            protected_memory[event_count][1] = event_priority;
            if (frame < 11)
                frame = (MAX_TEMP_MEMORY-1) - (frame_index+1);
            for (index=0; index<24; index++)
            {
                protected_memory[event_count][index+2] = temp_memory[frame][1];
                frame = (frame+1) % MAX_TEMP_MEMORY;
            }
            tx_mutex_put (&memory_mutex);
            event_count++;
        }
        else printf(" **not processed**");
        printf ("\n");
        tx_thread_suspend(&event_recorder);
    }
}
```

Figure 14.28: Function definitions Part 4—event_recorder entry function

We will not implement the overwrite rules in the case when protected memory is full. These rules depend on the actual file system used, and are left as an exercise for the reader. When protected memory is full, we will display a message to that effect, but will not copy the event.

Figure 14.29 contains the definition for the print_stats expiration function, which is part of the timer called stats_timer. This timer expires every 1,000 timer-ticks and the function displays a statistical summary of the system.

We have now discussed most of the components of the VAM system. The next section contains a complete listing of the system, which also appears on the CD at the back of the book.

```c
/***********************************************************/
/********** print statistics at specified times ***********/
/***********************************************************/
void print_stats (ULONG invalue)
{
    UINT row, col;
    printf("\n\n**** VAM System Periodic Event Summary\n\n");
    printf("      Current Time:                %lu\n", tx_time_get());
    printf("         Number of Crashes:        %lu\n", num_crashes);
    printf("         Number of Unsafe Events:  %lu\n", num_unsafe);
    printf("         Number of Warnings:       %lu\n", num_warning);
    printf("         Number of Manual Events:  %lu\n", num_manual);

    if (event_count > 0)
    {
        printf("\n\n**** Portion of Protected Memory Contents\n\n");
        printf("%6s%6s%6s\n", "Time", "Pri", "Data");
        for (row = 0; row < event_count; row++)
        {
            for (col = 0; col < 8; col++)
                printf("%6lu", protected_memory[row][col]);
            printf("      (etc.)\n");
        }
    }
    if (event_count >= MAX_EVENTS)
        printf("   Warning: Protected Memory is full...\n\n");
}
```

Figure 14.29: Function definitions Part 5—print_stats expiration function

14.6 Listing of VAM System

```
001  /* 14_case_study.c
002
003     Implement a simplified version of a real-time, video/audio/
           motion (VAM)
004     recording system.
005
006     Create three threads named: initializer, data_capture,
           event_recorder
007     Create one byte pool for thread stacks and message queue:
           my_byte_pool
008     Create one mutex to guard protected memory: memory_mutex
009     Create one message queue to store event notices:
           event_notice
010     Create nine application timers named: crash_interrupt,
           unsafe_interrupt,
011     warning_interrupt, manual_interrupt, crash_copy_scheduler,
012     unsafe_copy_scheduler, manual_copy_scheduler, stats_timer
013
014     For this system, assume that each timer-tick represents
           one second */
015
016  /**************************************************/
017  /*      Declarations, Definitions, and Prototypes     */
018  /**************************************************/
019
020  #include "tx_api.h"
021  #include <stdio.h>
022
023  #define STACK_SIZE      1024
024  #define BYTE_POOL_SIZE 9120
025  #define MAX_EVENTS        16
026  #define MAX_TEMP_MEMORY 200
027
028
029  /* Define the ThreadX object control blocks */
030
031  TX_THREAD     initializer;
032  TX_THREAD     data_capture;
```

```
033   TX_THREAD      event_recorder;
034
035   TX_QUEUE       event_notice;
036
037   TX_MUTEX       memory_mutex;
038   TX_BYTE_POOL my_byte_pool;
039
040   TX_TIMER       crash_interrupt;
041   TX_TIMER       unsafe_interrupt;
042   TX_TIMER       warning_interrupt;
043   TX_TIMER       manual_interrupt;
044
045   TX_TIMER       crash_copy_scheduler;
046   TX_TIMER       unsafe_copy_scheduler;
047   TX_TIMER       warning_copy_scheduler;
048   TX_TIMER       manual_copy_scheduler;
049   TX_TIMER       stats_timer;
050
051
052   /* Define the counters and variables used in the VAM system */
053
054   ULONG num_crashes=0, num_unsafe=0, num_warning=0,
         num_manual=0;
055   ULONG frame_index, event_count, frame_data[2];
056
057   /* Define the arrays used to represent temporary memory */
058   /* and protected memory. temp_memory contains pair of data */
059   /* in the form time-data and protected_memory contains rows */
060   /* of 26 elements in the form time-priority-data-data-data... */
061   /* The working index to temp_memory is frame_index and the */
062   /* working index to protected_memory is event_count. */
063
064   ULONG temp_memory[MAX_TEMP_MEMORY][2],
065         protected_memory[MAX_EVENTS][26];
066
067   /* Define thread and function prototypes. */
068
069   void initializer_process(ULONG);
070   void data_capture_process(ULONG);
071   void event_recorder_process(ULONG);
```

```
072   void crash_ISR(ULONG);
073   void unsafe_ISR(ULONG);
074   void warning_ISR(ULONG);
075   void manual_ISR(ULONG);
076   void crash_copy_activate(ULONG);
077   void unsafe_copy_activate(ULONG);
078   void warning_copy_activate(ULONG);
079   void manual_copy_activate(ULONG);
080   void print_stats(ULONG);
081
082
083   /****************************************************/
084   /*                  Main Entry Point                */
085   /****************************************************/
086
087   /* Define main entry point. */
088
089   int main()
090   {
091
092      /* Enter the ThreadX kernel. */
093      tx_kernel_enter();
094   }
095
096
097
098   /****************************************************/
099   /*              Application Definitions             */
100   /****************************************************/
101
102
103   /* Define what the initial system looks like. */
104
105   void tx_application_define(void *first_unused_memory)
106   {
107
108   CHAR *byte_pointer;
109
110     /* Put system definition stuff in here, e.g., thread creates
111        and other assorted create information. */
```

```
112
113     /* Create a memory byte pool from which to allocate
114        the thread stacks. */
115     tx_byte_pool_create(&my_byte_pool, "my_byte_pool",
116       first_unused_memory, BYTE_POOL_SIZE);
117
118     /* Allocate the stack for the initializer thread. */
119     tx_byte_allocate(&my_byte_pool, (VOID **) &byte_pointer,
120     STACK_SIZE, TX_NO_WAIT);
121
122     /* Create the initializer thread. */
123     tx_thread_create(&initializer, "initializer",
124                      initializer_process, 0,
125                      byte_pointer, STACK_SIZE, 11, 11,
126                      TX_NO_TIME_SLICE, TX_AUTO_START);
127
128     /* Allocate the stack for the data_capture thread. */
129     tx_byte_allocate(&my_byte_pool, (VOID **) &byte_pointer,
130       STACK_SIZE, TX_NO_WAIT);
131
132     /* Create the data_capture thread. */
133     tx_thread_create(&data_capture, "data_capture",
134                      data_capture_process, 0,
135                      byte_pointer, STACK_SIZE, 15, 15,
136                      TX_NO_TIME_SLICE, TX_AUTO_START);
137
138     /* Allocate the stack for the event_recorder thread. */
139     tx_byte_allocate(&my_byte_pool, (VOID **) &byte_pointer,
140       STACK_SIZE, TX_NO_WAIT);
141
142     /* Create the event_recorder thread. */
143     tx_thread_create(&event_recorder, "event_recorder",
144                      event_recorder_process, 0,
145                      byte_pointer, STACK_SIZE, 12, 12,
146                      TX_NO_TIME_SLICE, TX_DONT_START);
147
148     /* Create and activate the 4 timers to simulate interrupts */
149     tx_timer_create (&crash_interrupt, "crash_interrupt",
                         crash_ISR,
150                      0×1234, 1444, 1444, TX_AUTO_ACTIVATE);
```

```
151    tx_timer_create (&unsafe_interrupt, "unsafe_interrupt",
                         unsafe_ISR,
152                      0X1234, 760, 760, TX_AUTO_ACTIVATE);
153    tx_timer_create (&warning_interrupt, "warning_interrupt",
                         warning_ISR,
154                      0X1234, 410, 410, TX_AUTO_ACTIVATE);
155    tx_timer_create (&manual_interrupt, "manual_interrupt",
                         manual_ISR,
156                      0X1234, 888, 888, TX_AUTO_ACTIVATE);
157
158    /* Create and activate the 4 timers to initiate data
          copying */
159    tx_timer_create (&crash_copy_scheduler,
                         "crash_copy_scheduler",
160                      crash_copy_activate, 0X1234, 12, 12,
                         TX_NO_ACTIVATE);
161    tx_timer_create (&unsafe_copy_scheduler,
                         "unsafe_copy_scheduler",
162                      unsafe_copy_activate, 0X1234, 12, 12,
                         TX_NO_ACTIVATE);
163    tx_timer_create (&warning_copy_scheduler,
                         "warning_copy_scheduler",
164                      warning_copy_activate, 0X1234, 12, 12,
                         TX_NO_ACTIVATE);
165    tx_timer_create (&manual_copy_scheduler,
                         "manual_copy_scheduler",
166                      manual_copy_activate, 0X1234, 12, 12,
                         TX_NO_ACTIVATE);
167
168    /* Create and activate the timer to print statistics
          periodically */
169    tx_timer_create (&stats_timer, "stats_timer", print_stats,
170                      0X1234, 1000, 1000, TX_AUTO_ACTIVATE);
171
172    /* Create the message queue that holds the frame_indexes
          for all events. */
173    /* The frame_index is a position marker for the temp_memory
          array. */
174    /* Whenever an event occurs, the event ISR sends the
          current frame_index */
```

```
175    /* and event priority to the queue for storing crash event
          information. */
176    /* First, allocate memory space for the queue, then create
          the queue. */
177    tx_byte_allocate(&my_byte_pool, (VOID **) &byte_pointer,
178                    MAX_EVENTS*2*sizeof(ULONG), TX_NO_WAIT);
179    tx_queue_create (&event_notice, "event_notice", TX_2_ULONG,
180                    byte_pointer, MAX_EVENTS*2*sizeof(ULONG));
181
182    }
183
184
185    /**********************************************************/
186    /*                 Function Definitions                 */
187    /**********************************************************/
188
189
190    /* Entry function definition of the initializer thread */
191
192    void initializer_process(ULONG thread_input)
193    {
194    /* Perform initialization tasks */
195    /* Because we are using arrays to represent files, there is */
196    /* very little initialization to perform. We initialize two */
197    /* global variables that represent starting array indexes. */
198       printf("VAM System-Trace of Event Activities Begins…\n\n");
199       frame_index=0;
200       event_count=0;
201    }
202
203    /**********************************************************/
204    /* Entry function definition of the data_capture thread */
205
206    void data_capture_process(ULONG thread_input)
207    {
208    /* Perform data capture from the VAM system to temporary
          memory */
209    /* This function simulates the data capture operation by
          writing */
210    /* to an array, which represents a temporary memory file. For */
```

```
211    /* simplicity, we will write to the array once every timer-
          tick */
212       while (1) {
213         temp_memory[frame_index][0]=tx_time_get();
214         temp_memory[frame_index][1]=0X1234;
215         frame_index=(frame_index+1) % MAX_TEMP_MEMORY;
216         tx_thread_sleep(1);
217       }
218    }
219
220
221    /**********************************************************/
222    /********** crash event detection and processing **********/
223    /**********************************************************/
224
225    /* Timer function definition for simulated crash interrupt */
226    /* This is a simulated ISR -- an actual ISR would probably
          begin */
227    /* with a context save and would end with a context restore. */
228
229    void crash_ISR (ULONG timer_input)
230    {
231       ULONG frame_data[2];
232       frame_data[0]=frame_index;
233       frame_data[1]=1;
234       num_crashes++;
235    /* Activates timer to expire in 12 seconds-end of event */
236    /* Put frame_index and priority on queue for crash events */
237       tx_queue_send (&event_notice, frame_data, TX_NO_WAIT);
238       tx_timer_activate (&crash_copy_scheduler);
239    }
240    /**********************************************************/
241    /* Timer function definition for timer crash_copy_scheduler, */
242    /* which resumes thread that performs recording of crash data */
243
244    void crash_copy_activate (ULONG timer_input)
245    {
246       /* resume recorder thread to initiate data recording */
247       tx_thread_resume(&event_recorder);
248       tx_timer_deactivate (&crash_copy_scheduler);
249    }
```

```
250
251   /**********************************************************/
252   /**** Entry function definition of thread event_recorder ****/
253   /**********************************************************/
254
255   void event_recorder_process(ULONG thread_input)
256   {
257     ULONG frame, event_priority, event_time, index,
          frame_data[2];
258     while (1)
259     {
260     /* Copy an event from temporary memory to protected memory. */
261     /* Get frame_index from event message queue and copy 24
          frames */
262     /* from temp_memory to protected_memory. */
263       tx_queue_receive (&event_notice, frame_data, TX_NO_WAIT);
264       /* Store event time and event priority in protected
          memory */
265       frame=frame_data[0];
266       event_priority=frame_data[1];
267       event_time=temp_memory[frame][0];
268       printf("**Event** Time: %5lu Count: %2lu Pri: %lu",
269               event_time, event_count, event_priority);
270       if (event_count<MAX_EVENTS)
271       {
272         tx_mutex_get (&memory_mutex, TX_WAIT_FOREVER);
273         protected_memory[event_count][0]=event_time;
274         protected_memory[event_count][1]=event_priority;
275         if (frame<11)
276           frame=(MAX_TEMP_MEMORY-1)-(frame_index+1);
277         for (index=0; index<24; index++)
278         {
279           protected_memory[event_count][index+2]=temp_
              memory[frame][1];
280           frame=(frame+1) % MAX_TEMP_MEMORY;
281         }
282         tx_mutex_put (&memory_mutex);
283         event_count++;
284       }
285       else printf(" **not processed**");
```

```
286         printf ("\n");
287         tx_thread_suspend(&event_recorder);
288      }
289 }
290
291 /***********************************************************/
292 /********** unsafe event detection and processing **********/
293 /***********************************************************/
294
295 /* Timer function definition for simulated unsafe interrupt */
296 /* This is a simulated ISR -- an actual ISR would probably
       begin */
297 /* with a context save and would end with a context restore. */
298
299 void unsafe_ISR (ULONG timer_input)
300 {
301     ULONG frame_data[2];
302     frame_data[0]=frame_index;
303     frame_data[1]=2;
304     num_unsafe++;
305 /* Activates timer to expire in 12 seconds-end of event */
306 /* Put frame_index and priority on queue for unsafe events */
307     tx_queue_send (&event_notice, frame_data, TX_NO_WAIT);
308     tx_timer_activate (&unsafe_copy_scheduler);
309 }
310 /***********************************************************/
311 /* Timer function definition for timer unsafe_copy_scheduler, */
312 /* which resumes thread that performs recording of unsafe
       data */
313
314 void unsafe_copy_activate (ULONG timer_input)
315 {
316     /* resume event_recorder thread to initiate data
           recording */
317     tx_thread_resume(&event_recorder);
318     tx_timer_deactivate (&unsafe_copy_scheduler);
319 }
320
321
```

```
322  /**********************************************************/
323  /********* warning event detection and processing *********/
324  /**********************************************************/
325
326  /* Timer function definition for simulated warning interrupt */
327  /* This is a simulated ISR -- an actual ISR would probably
        begin */
328  /* with a context save and would end with a context restore. */
329
330  void warning_ISR (ULONG timer_input)
331  {
332     ULONG frame_data[2];
333     frame_data[0]=frame_index;
334     frame_data[1]=3;
335     num_warning++;
336  /* Activates timer to expire in 12 seconds-end of event */
337  /* Put frame_index and priority on queue for warning events */
338     tx_queue_send (&event_notice, frame_data, TX_NO_WAIT);
339     tx_timer_activate (&warning_copy_scheduler);
340  }
341  /**********************************************************/
342  /* Timer function definition for timer warning_copy_scheduler, */
343  /* which resumes thread that performs recording of warning
        data */
344
345  void warning_copy_activate (ULONG timer_input)
346  {
347     /* resume event_recorder thread to initiate data recording */
348     tx_thread_resume(&event_recorder);
349     tx_timer_deactivate (&warning_copy_scheduler);
350  }
351
352
353  /**********************************************************/
354  /********* manual event detection and processing *********/
355  /**********************************************************/
356
357  /* Timer function definition for simulated manual interrupt */
358  /* This is a simulated ISR -- an actual ISR would probably
        begin */
```

```
359   /* with a context save and would end with a context restore. */
360
361   void manual_ISR (ULONG timer_input)
362   {
363       ULONG frame_data[2];
364       frame_data[0]=frame_index;
365       frame_data[1]=4;
366       num_manual++;
367   /* Activates timer to expire in 12 seconds-end of event */
368   /* Put frame_index and priority on queue for manual events */
369       tx_queue_send (&event_notice, frame_data, TX_NO_WAIT);
370       tx_timer_activate (&manual_copy_scheduler);
371   }
372   /***********************************************************/
373   /* Timer function definition for timer manual_copy_scheduler, */
374   /* which resumes thread that performs recording of manual
      data */
375
376   void manual_copy_activate (ULONG timer_input)
377   {
378       /* resume event_recorder thread to initiate data recording */
379       tx_thread_resume(&event_recorder);
380       tx_timer_deactivate (&manual_copy_scheduler);
381   }
382
383
384   /***********************************************************/
385   /*********** print statistics at specified times ***********/
386   /***********************************************************/
387
388   void print_stats (ULONG invalue)
389   {
390       UINT row, col;
391       printf("\n\n**** VAM System Periodic Event Summary\n\n");
392       printf(" Current Time: %lu\n", tx_time_get());
393       printf(" Number of Crashes: %lu\n", num_crashes);
394       printf(" Number of Unsafe Events: %lu\n", num_unsafe);
395       printf(" Number of Warnings: %lu\n", num_warning);
396       printf(" Number of Manual Events: %lu\n", num_manual);
```

```
397
398    if (event_count>0)
399    {
400        printf("\n\n**** Portion of Protected Memory
              Contents\n\n");
401        printf("%6s%6s%6s\n", "Time", "Pri", "Data");
402        for (row=0; row<event_count; row++)
403        {
404            for (col=0; col<8; col++)
405                    printf("%6lu", protected_memory[row][col]);
406            printf(" (etc.)\n");
407        }
408    }
409    if (event_count >= MAX_EVENTS)
410        printf(" Warning: Protected Memory is full...\n\n");
411  }
```

14.7 Overview

This case study provides an excellent overview of developing a system with ThreadX. We used a variety of ThreadX services, including the following:

- application timers

- threads

- message queue

- mutex

- memory byte pool

This case study depends heavily on the use of application timers. One reason for using so many timers is because we need to schedule the copying of data from the temporary memory to the protected memory whenever any one of four events occurs. Our design provides the ability to record several events within each 24-second time frame, rather than just one. Application timers play a major role in providing this feature. We also used application timers to simulate interrupts that signify the occurrence of events, and we used one timer to display periodic system statistics.

We used only three threads for this case study. The *initializer* thread initializes the system, the *data_capture* thread coordinates the capture of data from the VAM unit to temporary memory, and the *event_recorder* thread copies event data from temporary memory to protected memory. We created the *data_capture* thread with the TX_AUTO_START option, so it would become ready immediately after initialization. We created the *event_recorder* thread with the TX_DON'T_START option, which means it cannot start until it is resumed by an application timer. When the *event_recorder* thread completes its data copying operation, it suspends itself and remains suspended until it is resumed by a timer.

We used one message queue to store information about events that had occurred. When the *event_recorder* thread is resumed, it takes the first message from the front of the message queue. This message contains information about the event data being copied from temporary memory to protected memory. Although we used only one message queue in this case study, typical applications tend to use a number of message queues.

We used one mutex to ensure that the *event_recorder* thread had exclusive access to protected memory before beginning the copying operation.

We used one memory byte pool to provide space for the thread stacks and for the message queue. This application is an excellent use of a memory byte pool because we need memory space of different sizes for the thread stacks and the message queue. Furthermore, these entities remain in existence for the life of the system, so there is no possibility of the fragmentation problems that can occur when bytes are repeatedly allocated and released.

We also demonstrated several examples of communication between resources. For example, timers send data to the message queue, a thread receives data from the message queue, a timer activates another timer, a timer resumes a thread, and a thread obtains ownership of a mutex. All these entities interact to form an efficient and effective system.

Appendices

This series of appendices comprises a reference manual for ThreadX.[1] Every service is described here, including purpose, parameters, return values, and examples. Following is a list of the services discussed in the appendices. Appendix K contains the ThreadX API, which summarizes all the service calls.

A   Memory Block Pool Services

B   Memory Byte Pool Services

C   Event Flags Group Services

D   Interrupt Control Service

E   Mutex Services

F   Message Queue Services

G   Counting Semaphore Services

H   Thread Services

I   Internal System Clock Services

J   Application Timer Services

K   ThreadX API

---

[1]ThreadX is a registered trademark of Express Logic, Inc. The ThreadX API, associated data structures, and data types are copyrights of Express Logic, Inc.

Appendices A through J follow a common format, as follows:

Service	Service name and brief description
Prototype	Prototype showing service name, parameter order, and data types
Description	**Detailed information about the service**
Input Parameters	Narne and description of parameters that provide data to the service (Note that the parameters may not be presented in the same order as the prototype, which specifies the correct order.)
Output Parameters	Name and description of parameters whose values are returned by the service
Return Values	Description of status values returned after invoking the service
Allowed From	List of objects that can call this service
Preemption Possible	Determination (Yes or No) of whether a preemption condition can arise as a result of calling this service, e.g., if "Yes," the thread executing this service may be preempted by a higher-priority thread because that service may resume a higher-priority thread
Example	Complete example using this service
See Also	List of related services

Memory Block Pool Services

The memory block pool services described in this appendix are:

tx_block_allocate	Allocate a fixed-size block of memory
tx_block_pool_create	Create a pool of fixed-size memory blocks
tx_block_pool_delete	Delete pool of fixed-size memory blocks
tx_block_pool_info_get	Retrieve information about block pool
tx_block_pool_performance_info_get	Get block pool performance information
tx_block_pool_performance_system_ info_get	Get block pool system performance information
tx_block_pool_prioritize	Prioritize block pool suspension list
tx_block_release	Release a fixed-size block of memory

tx_block_allocate

Allocate a fixed-size block of memory

Prototype

```
UINT tx_block_allocate (TX_BLOCK_POOL *pool_ptr, VOID **block_ptr,
                        ULONG wait_option)
```

Description

This service allocates a fixed-size memory block from the specified memory pool. The actual size of the memory block is determined during memory pool creation. This service modifies the Memory Block Pool Control Block through the parameter pool_ptr.

Input Parameter

pool_ptr   Pointer to a previously created pool's memory Control Block.

Output Parameters

block_ptr   Pointer to a destination block pointer. On successful allocation, the address of the allocated memory block is placed where this parameter points to.

wait_option   Defines how the service behaves if no memory blocks are available. The wait options are defined as follows:

TX_NO_WAIT (0x00000000)

TX_WAIT_FOREVER (0xFFFFFFFF)

timeout value (0x00000001 to 0xFFFFFFFE, inclusive)

Selecting TX_NO_WAIT results in an immediate return from this service regardless of whether or not it was successful. *This is the only valid option if the service is called from a non-thread; e.g., initialization, timer, or ISR.* Selecting TX_WAIT_FOREVER causes the calling thread to suspend indefinitely until a memory block becomes available. Selecting a numeric value (1-0xFFFFFFFE) specifies the maximum number of timer-ticks to stay suspended while waiting for a memory block.

Return Values

TX_SUCCESS[1]	(0x00)	Successful memory block allocation.
TX_DELETED	(0x01)	Memory block pool was deleted while thread was suspended.
TX_NO_MEMORY[1]	(0x10)	Service was unable to allocate a block of memory.

---

[1]This value is not affected by the TX_DISABLE_ERROR_CHECKING define that is used to disable API error checking.

TX_WAIT_ABORTED[1]   (0x1A)   Suspension was aborted by another thread, timer, or ISR.

TX_POOL_ERROR   (0x02)   Invalid memory block pool pointer.

TX_PTR_ERROR   (0x03)   Invalid pointer to destination pointer.

TX_WAIT_ERROR   (0x04)   A wait option other than TX_NO_WAIT was specified on a call from a non-thread.

Allowed From

Initialization, threads, timers, and ISRs

Preemption Possible

Yes

Example

```
TX_BLOCK_POOL my_pool;
unsigned char *memory_ptr;
UINT status;
...
/* Allocate a memory block from my_pool. Assume that the pool has
already been created with a call to tx_block_pool_create. */
status = tx_block_allocate (&my_pool, (VOID **) &memory_ptr,
  TX_NO_WAIT);
/* If status equals TX_SUCCESS, memory_ptr contains the address of
the allocated block of memory. */
```

tx_block_pool_create

Create a pool of fixed-size memory blocks

Prototype

```
UINT tx_block_pool_create (TX_BLOCK_POOL *pool_ptr,
  CHAR *name_ptr, ULONG block_size,
  VOID *pool_start, ULONG pool_size)
```

Description

This service creates a pool of fixed-size memory blocks. The memory area specified is divided into as many fixed-size memory blocks as possible using the formula:

```
total blocks = (total bytes) / (block size + size of (void*)).
```

This service initializes the Memory Block Pool Control Block through the parameter pool_ptr.

WARNING:

Each memory block contains one pointer of overhead that is invisible to the user and is represented by the "sizeof(void*)" expression in the preceding formula.

Input Parameters

pool_ptr	Pointer to a Memory Block Pool Control Block.
name_ptr	Pointer to the name of the memory block pool.
block_size	Number of bytes in each memory block.
pool_start	Starting address of the memory block pool.
pool_size	Total number of bytes available for the memory block pool.

Return Values

TX_SUCCESS[2]	(0x00)	Successful memory block pool creation.
TX_POOL_ERROR	(0x02)	Invalid memory block pool pointer. Either the pointer is NULL or the pool has already been created.
TX_PTR_ERROR	(0x03)	Invalid starting address of the pool.
TX_SIZE_ERROR	(0x05)	Size of pool is invalid.
TX_CALLER_ERROR	(0x13)	Invalid caller of this service.

---

[2]This value is not affected by the TX_DISABLE_ERROR_CHECKING define that is used to disable API error checking.

Allowed From

Initialization and threads

Preemption Possible

No

Example

```
TX_BLOCK_POOL my_pool;
UINT status;
...
/* Create a memory pool whose total size is 1000 bytes starting at
address 0x100000. Each block in this pool is defined to be 50 bytes
long. */
status = tx_block_pool_create (&my_pool, "my_pool_name",
  50, (VOID *) 0x100000, 1000);
/* If status equals TX_SUCCESS, my_pool contains 18 memory blocks
of 50 bytes each. The reason there are not 20 blocks in the pool is
because of the one overhead pointer associated with each block. */
```

tx_block_pool_delete

Delete a pool of fixed-size memory blocks

Prototype

```
UINT tx_block_pool_delete (TX_BLOCK_POOL *pool_ptr)
```

Description

This service deletes the specified block memory pool. All threads suspended waiting for a memory block from this pool are resumed and given a TX_DELETED return status.

WARNING:

It is the application's responsibility to manage the memory area associated with the pool, which is available after this service completes. In addition, the application must not use a deleted pool or its formerly allocated memory blocks.

Input Parameter

pool_ptr   Pointer to a previously created memory block pool's Control Block.

Return Values

TX_SUCCESS[3]	(0x00)	Successful memory block pool deletion.
TX_POOL_ERROR	(0x02)	Invalid memory block pool pointer.
TX_CALLER_ERROR	(0x13)	Invalid caller of this service.

Allowed From

Threads

Preemption Possible

Yes

Example

```
TX_BLOCK_POOL my_pool;
UINT status;
...
/* Delete entire memory block pool. Assume that the pool has
already been created with a call to tx_block_pool_create. */
status = tx_block_pool_delete (&my_pool);
/* If status equals TX_SUCCESS, the memory block pool is deleted.
*/
```

tx_block_pool_info_get

Retrieve information about a memory block pool

Prototype

```
UINT tx_block_pool_info_get(TX_BLOCK_POOL *pool_ptr, CHAR **name,
  ULONG *available, ULONG *total_blocks,
  TX_THREAD **first_suspended,
```

------

[3]This value is not affected by the TX_DISABLE_ERROR_CHECKING define that is used to disable API error checking.

```
ULONG *suspended_count,
TX_BLOCK_POOL **next_pool)
```

Description

This service retrieves information about the specified block memory pool.

Input Parameter

pool_ptr   Pointer to previously created memory block pool's Control Block.

Output Parameters

name	Pointer to destination for the pointer to the block pool's name.
available	Pointer to destination for the number of available blocks in the block pool.
total_blocks	Pointer to destination for the total number of blocks in the block pool.
first_suspended	Pointer to destination for the pointer to the thread that is first on the suspension list of this block pool.
suspended_count	Pointer to destination for the number of threads currently suspended on this block pool.
next_pool	Pointer to destination for the pointer of the next created block pool.

Return Values

TX_SUCCESS[4]	(0x00)	(0x00) Successful block pool information retrieve.
TX_POOL_ERROR	(0x02)	Invalid memory block pool pointer.
TX_PTR_ERROR	(0x03)	Invalid pointer (NULL) for any destination pointer.

[4]This value is not affected by the TX_DISABLE_ERROR_CHECKING define that is used to disable API error checking.

Allowed From

Initialization, threads, timers, and ISRs

Preemption Possible

No

Example

```
TX_BLOCK_POOL my_pool;
CHAR *name;
ULONG available;
ULONG total_blocks;
TX_THREAD *first_suspended;
ULONG suspended_count;
TX_BLOCK_POOL *next_pool;
UINT status;
...
/* Retrieve information about the previously created block pool
"my_pool." */
status = tx_block_pool_info_get(&my_pool, &name,
  &available,&total_blocks,
  &first_suspended, &suspended_count,
  &next_pool);
/* If status equals TX_SUCCESS, the information requested is
valid. */
```

tx_block_pool_performance_info_get

Get block pool performance information

Prototype

```
UINT tx_block_pool_performance_info_get(TX_BLOCK_POOL *pool_ptr,
  ULONG *allocates,
  ULONG *releases,
  ULONG *suspensions,
  ULONG *timeouts)
```

Description

This service retrieves performance information about the specified memory block pool.

> **NOTE:**
>
> The ThreadX library and application must be built with TX_BLOCK_POOL_ENABLE_PERFORMANCE_INFO defined for this service to return performance information.

Input Parameters

pool_ptr	Pointer to previously created memory block pool.
allocates	Pointer to destination for the number of allocate requests performed on this pool.
releases	Pointer to destination for the number of release requests performed on this pool.
suspensions	Pointer to destination for the number of thread allocation suspensions on this pool.
timeouts	Pointer to destination for the number of allocate suspension timeouts on this pool.

> **NOTE:**
>
> Supplying a TX_NULL for any parameter indicates that the parameter is not required.

Return Values

TX_SUCCESS (0x00)	Successful block pool performance get.
TX_PTR_ERROR (0x03)	Invalid block pool pointer.
TX_FEATURE_NOT_ENABLED (0xFF)	The system was not compiled with performance information enabled.

Allowed From

Initialization, threads, timers, and ISRs

Example

```
TX_BLOCK_POOL my_pool;
  ULONG allocates;
  ULONG releases;
  ULONG suspensions;
  ULONG timeouts;
…

/* Retrieve performance information on the previously created
block pool. */
status = tx_block_pool_performance_info_get(&my_pool, &allocates,
  &releases, &suspensions, &timeouts);
/* If status is TX_SUCCESS the performance information was
successfully retrieved. */
```

See Also

tx_block_allocate, tx_block_pool_create, tx_block_pool_delete, tx_block_pool_info_get, tx_block_pool_performance_info_get, tx_block_pool_performance_system_info_get, tx_block_release

tx_block_pool_performance_system_info_get

Get block pool system performance information

Prototype

```
UINT tx_block_pool_performance_system_info_get(ULONG *allocates,
  ULONG *releases,
  ULONG *suspensions,
  ULONG *timeouts);
```

Description

This service retrieves performance information about all memory block pools in the application.

> **NOTE:**
>
> The ThreadX library and application must be built with TX_BLOCK_POOL_ENABLE_
> PERFORMANCE_INFO defined for this service to return performance information.

Input Parameters

allocates	Pointer to destination for the total number of allocate requests performed on all block pools.
releases	Pointer to destination for the total number of release requests performed on all block pools.
suspensions	Pointer to destination for the total number of thread allocation suspensions on all block pools.
timeouts	Pointer to destination for the total number of allocate suspension timeouts on all block pools.

> **NOTE:**
>
> Supplying a TX_NULL for any parameter indicates that the parameter is not required.

Return Values

TX_SUCCESS (0x00)	Successful block pool system performance get.
TX_FEATURE_NOT_ENABLED (0xFF)	The system was not compiled with performance information enabled.

Allowed From

Initialization, threads, timers, and ISRs

Example

```
ULONG allocates;
ULONG releases;
ULONG suspensions;
```

```
ULONG timeouts;
...
/* Retrieve performance information on all the block pools in the
system. */
status = tx_block_pool_performance_system_info_get(&allocates,
  &releases,
  &suspensions,
  &timeouts);
/* If status is TX_SUCCESS the performance information was
successfully retrieved. */
```

See Also

tx_block_allocate, tx_block_pool_create, tx_block_pool_delete, tx_block_pool_info_get, tx_block_pool_performance_info_get, tx_block_pool_prioritize, tx_block_release

tx_block_pool_prioritize

Prioritize the memory block pool suspension list

Prototype

```
UINT tx_block_pool_prioritize(TX_BLOCK_POOL *pool_ptr)
```

Description

This service places the highest-priority thread suspended for a block of memory on this pool at the front of the suspension list. All other threads remain in the same FIFO order in which they were suspended.

Input Parameter

pool_ptr Pointer to a previously created memory block pool's Control Block.

Return Values

TX_SUCCESS[5]	(0x00)	Successful block pool prioritize.
TX_POOL_ERROR	(0x02)	Invalid memory block pool pointer.

---

[5]This value is not affected by the TX_DISABLE_ERROR_CHECKING define that is used to disable API error checking.

Allowed From

Initialization, threads, timers, and ISRs

Preemption Possible

No

Example

```
TX_BLOCK_POOL my_pool;
UINT status;
...
/* Ensure that the highest priority thread will receive the next
free block in this pool. */
status = tx_block_pool_prioritize(&my_pool);
/* If status equals TX_SUCCESS, the highest priority suspended
thread is at the front of the list. The next tx_block_release call
will wake up this thread. */
```

tx_block_pool_release

Release a fixed-size block of memory

Prototype

```
UINT tx_block_release(VOID *block_ptr)
```

Description

This service releases a previously allocated block back to its associated memory pool. If one or more threads are suspended waiting for a memory block from this pool, the first thread on the suspended list is given this memory block and resumed.

WARNING:

The application must not use a memory block area after it has been released back to the pool.

Input Parameter

block_ptr   Pointer to the previously allocated memory block.

Return Values

TX_SUCCESS[6]	(0x00)	Successful memory block release.
TX_PTR_ERROR	(0x03)	Invalid pointer to memory block.

Allowed From

Initialization, threads, timers, and ISRs

Preemption Possible

Yes

Example

```
TX_BLOCK_POOL my_pool;
unsigned char *memory_ptr;
UINT status;
…
/* Release a memory block back to my_pool. Assume that the pool
has been created and the memory block has been allocated. */
status = tx_block_release((VOID *) memory_ptr);
/* If status equals TX_SUCCESS, the block of memory pointed to by
memory_ptr has been returned to the pool. */
```

---

[6]This value is not affected by the TX_DISABLE_ERROR_CHECKING define that is used to disable API error checking.

Memory Byte Pool Services

The memory byte pool services described in this appendix are:

tx_byte_allocate	Allocate bytes of memory
tx_byte_pool_create	Create a memory pool of bytes
tx_byte_pool_delete	Delete a memory pool of bytes
tx_byte_pool_info_get	Retrieve information about a byte pool
tx_byte_pool_performance_info_get	Get byte pool performance information
tx_byte_pool_performance_system_info_get	Get byte pool system performance information
tx_byte_pool_prioritize	Prioritize the byte pool suspension list
tx_byte_release	Release bytes back to the memory pool

tx_byte_allocate

Allocate bytes of memory from a memory byte pool

Prototype

```
UINT tx_byte_allocate(TX_BYTE_POOL *pool_ptr,
                      VOID **memory_ptr, ULONG memory_size,
                      ULONG wait_option)
```

Description

This service allocates the specified number of bytes from the specified byte memory pool. This service modifies the Memory Pool Control Block through the parameter pool_ptr.

WARNING:

The performance of this service is a function of the block size and the amount of fragmentation in the pool. Hence, this service should not be used during time-critical threads of execution.

Input Parameters

pool_ptr	Pointer to a previously created memory byte pool's Control Block.
memory_size	Number of bytes requested.
wait_option	Defines how the service behaves if there is not enough memory available. The wait options are defined as follows:
	TX_NO_WAIT (0x00000000)
	TX_WAIT_FOREVER (0xFFFFFFFF)
	timeout value (0x00000001 to 0xFFFFFFFE, inclusive)
	Selecting TX_NO_WAIT results in an immediate return from this service regardless of whether or not it was successful. *This is the only valid option if the service is called from initialization.* Selecting TX_WAIT_FOREVER causes the calling thread to suspend indefinitely until enough memory is available. Selecting a numeric value (1-0xFFFFFFFE) specifies the maximum number of timer-ticks to stay suspended while waiting for the memory.

Output Parameter

memory_ptr	Pointer to a destination memory pointer. On successful allocation, the address of the allocated memory area is placed where this parameter points to.

Return Values

TX_SUCCESS[1]	(0x00)	Successful memory allocation.
TX_DELETED[1]	(0x01)	Memory pool was deleted while thread was suspended.
TX_NO_MEMORY[1]	(0x10)	Service was unable to allocate the memory.
TX_WAIT_ABORTED[1]	(0x1A)	Suspension was aborted by another thread, timer, or ISR.
TX_POOL_ERROR	(0x02)	Invalid memory pool pointer.
TX_PTR_ERROR	(0x03)	Invalid pointer to destination pointer.
TX_WAIT_ERROR	(0x04)	A wait option other than TX_NO_WAIT was specified on a call from a non-thread.
TX_CALLER_ERROR	(0x13)	Invalid caller of this service.

Allowed From

Initialization and threads

Preemption Possible

Yes

Example

```
TX_BYTE_POOL my_pool;
unsigned char *memory_ptr;
UINT status;
...
/* Allocate a 112 byte memory area from my_pool. Assume that the
pool has already been created with a call to tx_byte_pool_create. */

  status = tx_byte_allocate(&my_pool, (VOID **) &memory_ptr,

                    112, TX_NO_WAIT);
```

---

[1]This value is not affected by the TX_DISABLE_ERROR_CHECKING define that is used to disable API error checking.

```
/* If status equals TX_SUCCESS, memory_ptr contains the address of
the allocated memory area. */
```

tx_byte_pool_create

Create a memory pool of bytes

Prototype

```
UINT tx_byte_pool_create(TX_BYTE_POOL *pool_ptr,
                         CHAR *name_ptr, VOID *pool_start,
                         ULONG pool_size)
```

Description

This service creates a memory pool in the area specified. Initially, the pool consists of basically one very large free block. However, the pool is broken into smaller blocks as allocations are performed. This service initializes the Memory Pool Control Block through the parameter pool_ptr.

Input Parameters

pool_ptr	Pointer to a Memory Pool Control Block.
name_ptr	Pointer to the name of the memory pool.
pool_start	Starting address of the memory pool.
pool_size	Total number of bytes available for the memory pool.

Return Values[2]

TX_SUCCESS[2]	(0x00)	Successful memory pool creation.
TX_POOL_ERROR	(0x02)	Invalid memory pool pointer. Either the pointer is

---

[2]This value is not affected by the TX_DISABLE_ERROR_CHECKING define that is used to disable API error checking.

		NULL or the pool has already been created.
TX_PTR_ERROR	(0x03)	Invalid starting address of the pool.
TX_SIZE_ERROR	(0x05)	Size of pool is invalid.
TX_CALLER_ERROR	(0x13)	Invalid caller of this service.

Allowed From

Initialization and threads

Preemption Possible

No

Example

```
TX_BYTE_POOL my_pool;
UINT status;
/* Create a memory pool whose total size is 2000 bytes starting at
address 0x500000. */
status = tx_byte_pool_create(&my_pool, "my_pool_name",
                            (VOID *) 0x500000, 2000);
/* If status equals TX_SUCCESS, my_pool is available for
allocating memory. */
```

tx_byte_pool_delete

Delete a memory pool of bytes

Prototype

```
UINT tx_byte_pool_delete(TX_BYTE_POOL *pool_ptr)
```

Description

This service deletes the specified memory pool. All threads suspended waiting for memory from this pool are resumed and receive a TX_DELETED return status.

> **WARNING:**
>
> It is the application's responsibility to manage the memory area associated with the pool, which is available after this service completes. In addition, the application must not use a deleted pool or memory previously allocated from it.

Input Parameter

pool_ptr   Pointer to a previously created memory pool's Control Block.

Return Values[3]

TX_SUCCESS[3]	(0x00)	Successful memory pool deletion.
TX_POOL_ERROR	(0x02)	Invalid memory pool pointer.
TX_CALLER_ERROR	(0x13)	Invalid caller of this service.

Allowed From

Threads
Preemption Possible
Yes

Example

```
TX_BYTE_POOL my_pool;
UINT status;
...
/* Delete entire memory pool. Assume that the pool has already
been created with a call to tx_byte_pool_create. */
status = tx_byte_pool_delete(&my_pool);
/* If status equals TX_SUCCESS, memory pool is deleted. */
```

tx_byte_pool_info_get

Retrieve information about a memory byte pool

---

[3]This value is not affected by the TX_DISABLE_ERROR_CHECKING define that is used to disable API error checking.

Prototype

```
UINT tx_byte_pool_info_get(TX_BYTE_POOL *pool_ptr, CHAR **name,
                           ULONG *available, ULONG *fragments,
                           TX_THREAD **first_suspended,
                           ULONG *suspended_count,
                           TX_BYTE_POOL **next_pool)
```

Description

This service retrieves information about the specified memory byte pool.

Input Parameter

pool_ptr   Pointer to a previously created memory byte pool's Control Block.

Output Parameters

name	Pointer to destination for the pointer to the byte pool's name.
available	Pointer to destination for the number of available bytes in the pool.
fragments	Pointer to destination for the total number of memory fragments in the byte pool.
first_suspended	Pointer to destination for the pointer to the thread that is first on the suspension list of this byte pool.
suspended_count	Pointer to destination for the number of threads currently suspended on this byte pool.
next_pool	Pointer to destination for the pointer of the next created byte pool.

Return Values

TX_SUCCESS[4]	(0x00)	Successful pool information retrieval.
TX_POOL_ERROR	(0x02)	Invalid memory pool pointer.
TX_PTR_ERROR	(0x03)	Invalid pointer (NULL) for any destination pointer.

---

[4]This value is not affected by the TX_DISABLE_ERROR_CHECKING define that is used to disable API error checking.

Allowed From

Initialization, threads, timers, and ISRs

Preemption Possible

No

Example

```
TX_BYTE_POOL my_pool;
CHAR *name;
ULONG available;
ULONG fragments;
TX_THREAD *first_suspended;
ULONG suspended_count;
TX_BYTE_POOL *next_pool;
UINT status;
...
/* Retrieve information about the previously created block pool
"my_pool." */
status = tx_byte_pool_info_get(&my_pool, &name, &available,
                              &fragments,
                              &first_suspended, &suspended_count,
                              &next_pool);
/* If status equals TX_SUCCESS, the information requested is
valid. */
```

tx_byte_pool_performance_info_get

Get byte pool performance information

Prototype

```
UINT tx_byte_pool_performance_info_get(TX_BYTE_POOL *pool_ptr,

  ULONG *allocates, ULONG *releases,
  ULONG *fragments_searched,
  ULONG *merges, ULONG *splits,
  ULONG *suspensions,
  ULONG *timeouts);
```

Description

This service retrieves performance information about the specified memory byte pool.

NOTE:

The ThreadX library and application must be built with TX_BYTE_POOL_ENABLE_ PERFORMANCE_INFO defined for this service to return performance information.

Input Parameters

pool_ptr	Pointer to previously created memory byte pool.
allocates	Pointer to destination for the number of allocate requests performed on this pool.
releases	Pointer to destination for the number of release requests performed on this pool.
fragments_ searched	Pointer to destination for the number of internal memory fragments searched during allocation requests on this pool.
merges	Pointer to destination for the number of internal memory blocks merged during allocation requests on this pool.
splits	Pointer to destination for the number of internal memory blocks split (fragments) created during allocation requests on this pool.
suspensions	Pointer to destination for the number of thread allocation suspensions on this pool.
timeouts	Pointer to destination for the number of allocate suspension timeouts on this pool.

Return Values

TX_SUCCESS (0x00)	Successful byte pool performance get.
TX_PTR_ERROR (0x03)	Invalid byte pool pointer.
TX_FEATURE_NOT_ENABLED (0xFF)	The system was not compiled with performance information enabled.

Allowed From

Initialization, threads, timers, and ISRs

Example

```
TX_BYTE_POOL my_pool;
ULONG fragments_searched;
ULONG merges;
ULONG splits;
ULONG allocates;
ULONG releases;
ULONG suspensions;
ULONG timeouts;
...
/* Retrieve performance information on the previously created byte
pool. */
status = tx_byte_pool_performance_info_get
(&my_pool, &fragments_searched,
 &merges, &splits, &allocates,
 &releases, &suspensions,
 &timeouts);
/* If status is TX_SUCCESS the performance information was
successfully retrieved. */
```

See Also

tx_byte_allocate, tx_byte_pool_create, tx_byte_pool_delete, tx_byte_pool_info_get, tx_byte_pool_performance_system_info_get, tx_byte_pool_prioritize, tx_byte_release

tx_byte_pool_performance_system_info_get

Get byte pool system performance information

Prototype

```
UINT tx_byte_pool_performance_system_info_get(ULONG *allocates,
  ULONG *releases,
  ULONG *fragments_searched,
```

```
ULONG *merges,
ULONG *splits,
ULONG *suspensions,
ULONG *timeouts);
```

Description

This service retrieves performance information about all memory byte pools in the system.

NOTE:

The ThreadX library and application must be built with TX_BYTE_POOL_ENABLE_ PERFORMANCE_INFO defined for this service to return performance information.

Input Parameters

allocates	Pointer to destination for the number of allocate requests performed on this pool.
releases	Pointer to destination for the number of release requests performed on this pool.
fragments_ searched	Pointer to destination for the total number of internal memory fragments searched during allocation requests on all byte pools.
merges	Pointer to destination for the total number of internal memory blocks merged during allocation requests on all byte pools.
splits	Pointer to destination for the total number of internal memory blocks split (fragments) created during allocation requests on all byte pools.
suspensions	Pointer to destination for the total number of thread allocation suspensions on all byte pools.
timeouts	Pointer to destination for the total number of allocate suspension timeouts on all byte pools.

NOTE:

Supplying a TX_NULL for any parameter indicates the parameter is not required.

Return Values

TX_SUCCESS (0x00)	Successful byte pool performance get.
TX_FEATURE_NOT_ENABLED (0xFF)	The system was not compiled with performance information enabled.

Allowed From

Initialization, threads, timers, and ISRs

Example

```
ULONG fragments_searched;
ULONG merges;
ULONG splits;
ULONG allocates;
ULONG releases;
ULONG suspensions;
ULONG timeouts;
...
/* Retrieve performance information on all byte pools in the
system. */
status = tx_byte_pool_performance_system_info_get(
 &fragments_searched,
 &merges, &splits,
 &allocates, &releases,
 &suspensions, &timeouts);
/* If status is TX_SUCCESS the performance information was
successfully retrieved. */
```

See Also

tx_byte_allocate, tx_byte_pool_create, tx_byte_pool_delete, tx_byte_pool_info_get, tx_byte_pool_performance_info_get, tx_byte_pool_prioritize, tx_byte_release

tx_byte_pool_prioritize

Prioritize the memory byte pool suspension list

Prototype

```
UINT tx_byte_pool_prioritize(TX_BYTE_POOL *pool_ptr)
```

Description

This service places the highest-priority thread suspended for memory on this pool at the front of the suspension list. All other threads remain in the same FIFO order in which they were suspended.

Input Parameter

pool_ptr   Pointer to a previously created memory pool's Control Block.

Return Values[5]

TX_SUCCESS[5]	(0x00)	Successful memory pool prioritize.
TX_POOL_ERROR	(0x02)	Invalid memory pool pointer.

Allowed From

Initialization, threads, timers, and ISRs

Preemption Possible

No

Example

```
TX_BYTE_POOL my_pool;
UINT status;
…
/* Ensure that the highest priority thread will receive the next
   free memory from this pool. */
status = tx_byte_pool_prioritize(&my_pool);
/* If status equals TX_SUCCESS, the highest priority suspended
   thread is at the front of the list. The next tx_byte_release
```

---

[5]This value is not affected by the TX_DISABLE_ERROR_CHECKING define that is used to disable API error checking.

```
call will wake up this thread, if there is enough memory to
satisfy its request. */
```

tx_byte_release

Release bytes back to a memory byte pool

Prototype

```
UINT tx_byte_release(VOID *memory_ptr)
```

Description

This service releases a previously allocated memory area back to its associated pool. If one or more threads are suspended waiting for memory from this pool, each suspended thread is given memory and resumed until the memory is exhausted or until there are no more suspended threads. This process of allocating memory to suspended threads always begins with the first thread on the suspended list.

WARNING:

The application must not use the memory area after it is released.

Input Parameter

memory_ptr   Pointer to the previously allocated memory area.

Return Values[6]

TX_SUCCESS[6]	(0x00)	Successful memory release.
TX_PTR_ERROR	(0x03)	Invalid memory area pointer.
TX_CALLER_ERROR	(0x13)	Invalid caller of this service.

Allowed From

Initialization and threads

---

[6]This value is not affected by the TX_DISABLE_ERROR_CHECKING define that is used to disable API error checking.

Preemption Possible

Yes

Example

```
unsigned char *memory_ptr;
UINT status;
...
/* Release a memory back to my_pool. Assume that the memory area
was previously allocated from my_pool. */
status = tx_byte_release((VOID *) memory_ptr);
/* If status equals TX_SUCCESS, the memory pointed to by
memory_ptr has been returned to the pool. */
```

Event Flags Group Services

The event flags group services described in this appendix are:

tx_event_flags_create	Create an event flags group
tx_event_flags_delete	Delete an event flags group
tx_event_flags_get	Get event flags from an event flags group
tx_event_flags_info_get	Retrieve information about an event flags group
tx_event_flags_performance info_get	Get event flags group performance information
tx_event_flags_performance_system_info_get	Retrieve performance system information
tx_event_flags_set	Set event flags in an event flags group
tx_event_flags_set_notify	Notify application when event flags are set

tx_event_flags_create

Create an event flags group

Prototype

```
UINT tx_event_flags_create (TX_EVENT_FLAGS_GROUP *group_ptr,
                            CHAR *name_ptr)
```

Description

This service creates a group of 32 event flags. All 32 event flags in the group are initialized to zero. Each event flag is represented by a single bit. This service initializes the group Control Block through the parameter group_ptr.

Input Parameters

name_ptr   Pointer to the name of the event flags group.
group_ptr   Pointer to an Event Flags Group Control Block.

Return Values

TX_SUCCESS1   (0×00)   Successful event group creation.
TX_GROUP_ERROR   (0×06)   Invalid event group pointer. Either the
                                  pointer is NULL or the event group has
                                  already been created.
TX_CALLER_ERROR   (0×13)   Invalid caller of this service.

Allowed From

Initialization and threads

Preemption Possible

No

Example

```
TX_EVENT_FLAGS_GROUP my_event_group;
UINT status;
…
/* Create an event flags group. */
```

---

[1]This value is not affected by the TX_DISABLE_ERROR_CHECKING define that is used to disable API error checking.

```
status = tx_event_flags_create (&my_event_group,
                                "my_event_group_name");
/* If status equals TX_SUCCESS, my_event_group is ready for get
and set services. */
```

tx_event_flags_delete

Delete an event flags group

Prototype

```
UINT tx_event_flags_delete (TX_EVENT_FLAGS_GROUP *group_ptr)
```

Description

This service deletes the specified event flags group. All threads suspended waiting for events from this group are resumed and receive a TX_DELETED return status.

> **WARNING:**
>
> The application must not use a deleted event flags group.

Input Parameter

group_ptr   Pointer to a previously created event flags group's Control Block.

Return Values

TX_SUCCESS[2]	(0×00)	Successful event flags group deletion.
TX_GROUP_ERROR	(0×06)	Invalid event flags group pointer.
TX_CALLER_ERROR	(0×13)	Invalid caller of this service.

Allowed From

Threads

Preemption Possible

Yes

---

[2]This value is not affected by the TX_DISABLE_ERROR_CHECKING define that is used to disable API error checking.

Example

```
TX_EVENT_FLAGS_GROUP my_event_group;
UINT status;
...
/* Delete event flags group. Assume that the group has already been
created with a call to tx_event_flags_create. */
status = tx_event_flags_delete (&my_event_group);
/* If status equals TX_SUCCESS, the event flags group is deleted.
*/
```

tx_event_flags_get

Get event flags from an event flags group

Prototype

```
UINT tx_event_flags_get (TX_EVENT_FLAGS_GROUP *group_ptr,

    ULONG requested_flags, UINT get_option,
    ULONG *actual_flags_ptr, ULONG wait_option)
```

Description

This service waits on event flags from the specified event flags group. Each event flags group contains 32 event flags. Each flag is represented by a single bit. This service can wait on a variety of event flag combinations, as selected by the parameters. If the requested combination of flags is not set, this service either returns immediately, suspends until the request is satisfied, or suspends until a time-out is reached, depending on the wait option specified.

Input Parameters

group_ptr	Pointer to a previously created event flags group's Control Block.
requested_flags	32-bit unsigned variable that represents the requested event flags.

get_option
Specifies whether all or any of the requested event flags are required. The following are valid selections:
TX_AND (0×02)
TX_AND_CLEAR (0×03)
TX_OR (0×00)
TX_OR_CLEAR (0×01)
Selecting TX_AND or TX_AND_CLEAR specifies that all event flags must be set (a logical '1') within the group. Selecting TX_OR or TX_OR_CLEAR specifies that any event flag is satisfactory. Event flags that satisfy the request are cleared (set to zero) if TX_AND_CLEAR or TX_OR_CLEAR are specified.

wait_option
Defines how the service behaves if the selected event flags are not set. The wait options are defined as follows:
TX_NO_WAIT (0×00000000)
TX_WAIT_FOREVER (0×FFFFFFFF)
timeout value (0x00000001 to 0×FFFFFFFE, inclusive)
Selecting TX_NO_WAIT results in an immediate return from this service regardless of whether or not it was successful. This is the only valid option if the service is called from a non-thread; e.g., initialization, timer, or ISR. Selecting TX_WAIT_FOREVER causes the calling thread to suspend indefinitely until the event flags are available. Selecting a numeric value (1-0×FFFFFFFE) specifies the maximum number of timer-ticks to stay suspended while waiting for the event flags.

Output Parameter

actual_flags_ptr
Pointer to destination where the retrieved event flags are placed. Note that the actual flags obtained may contain flags that were not requested.

Return Values

TX_SUCCESS[3]	(0×00)	Successful event flags get.
TX_DELETED[3]	(0×01)	Event flags group was deleted while thread was suspended.
TX_NO_EVENTS[3]	(0×07)	Service was unable to get the specified events.
TX_WAIT_ABORTED[3]	(0×1A)	Suspension was aborted by another thread, timer, or ISR.
TX_GROUP_ERROR	(0×06)	Invalid event flags group pointer.
TX_PTR_ERROR	(0×03)	Invalid pointer for actual event flags.
TX_WAIT_ERROR	(0×04)	A wait option other than TX_NO_WAIT was specified on a call from a non-thread.
TX_OPTION_ERROR	(0×08)	Invalid get-option was specified.

Allowed From

Initialization, threads, timers, and ISRs

Preemption Possible

Yes

Example

```
TX_EVENT_FLAGS_GROUP my_event_group;
ULONG actual_events;
UINT status;
...
/* Request that event flags 0, 4, and 8 are all set. Also, if they
are set they should be cleared. If the event flags are not set,
this service suspends for a maximum of 20 timer-ticks. */

status = tx_event_flags_get(&my_event_group, 0x111,
                    TX_AND_CLEAR, &actual_events, 20);
```

---

[3]This value is not affected by the TX_DISABLE_ERROR_CHECKING define that is used to disable API error checking.

```
/* If status equals TX_SUCCESS, actual_events contains the actual
events obtained. */
```

tx_event_flags_info_get

Retrieve information about an event flags group.

Prototype

```
UINT tx_event_flags_info_get (TX_EVENT_FLAGS_GROUP *group_ptr,
   CHAR **name, ULONG *current_flags,
   TX_THREAD **first_suspended,
   ULONG *suspended_count,
   TX_EVENT_FLAGS_GROUP **next_group)
```

Description

This service retrieves information about the specified event flags group.

Input Parameter

group_ptr   Pointer to an Event Flags Group Control Block.

Output Parameters

name   Pointer to destination for the pointer to the event flags
                            group's name.
current_flags   Pointer to destination for the current set flags in the event
                            flags group.
first_suspended   Pointer to destination for the pointer to the thread that is
                            first on the suspension list of this event flags group.
suspended_count   Pointer to destination for the number of threads currently
                            suspended on this event flags group.
next_group   Pointer to destination for the pointer of the next created
                            event flags group.

Return Values

TX_SUCCESS[4]	(0 × 00)	Successful event group information retrieval.
TX_GROUP_ERROR	(0 × 06)	Invalid event group pointer.
TX_PTR_ERROR	(0 × 03)	Invalid pointer (NULL) for any destination pointer.

Allowed From

Initialization, threads, timers, and ISRs

Preemption Possible

No

Example

```
TX_EVENT_FLAGS_GROUP my_event_group;
CHAR *name;
ULONG current_flags;
TX_THREAD *first_suspended;
ULONG suspended_count;
TX_EVENT_FLAGS_GROUP *next_group;
UINT status;
...
/* Retrieve information about the previously created event flags
group "my_event_group." */
status = tx_event_flags_info_get(&my_event_group, &name,
                                 &current_flags,
                                 &first_suspended, &suspended_count,
                                 &next_group);
/* If status equals TX_SUCCESS, the information requested is
valid. */
```

tx_event_flags_performance info_get

Get event flags group performance information

---

[4]This value is not affected by the TX_DISABLE_ERROR_CHECKING define that is used to disable API error checking.

Prototype

```
UINT tx_event_flags_performance_info_get

  (TX_EVENT_FLAGS_GROUP *group_ptr,
  ULONG *sets, ULONG *gets,
  ULONG *suspensions, ULONG *timeouts);
```

Description

This service retrieves performance information about the specified event flags group.

> **NOTE:**
>
> ThreadX library and application must be built with TX_EVENT_FLAGS_ENABLE_PERFORMANCE_INFO defined for this service to return performance information.

Input Parameters

group_ptr	Pointer to previously created event flags group.
sets	Pointer to destination for the number of event flags set requests performed on this group.
gets	Pointer to destination for the number of event flags get requests performed on this group.
suspensions	Pointer to destination for the number of thread event flags get suspensions on this group.
timeouts	Pointer to destination for the number of event flags get suspension timeouts on this group.

> **NOTE:**
>
> Supplying a TX_NULL for any parameter indicates that the parameter is not required.

Return Values

TX_SUCCESS (0x00)	Successful event flags group performance get.
TX_PTR_ERROR (0x03)	Invalid event flags group pointer.
TX_FEATURE_NOT_ENABLED (0xFF)	The system was not compiled with performance information enabled.

Allowed From

Initialization, threads, timers, and ISRs

Example

```
TX_EVENT_FLAGS_GROUP my_event_flag_group;
ULONG sets;
ULONG gets;
ULONG suspensions;
ULONG timeouts;
...
/* Retrieve performance information on the previously created
event flag group. */
status = tx_event_flags_performance_info_get (&my_event_flag_group,
                                              &sets,
                                              &gets, &suspensions,
                                              &timeouts);
/* If status is TX_SUCCESS the performance information was
successfully retrieved. */
```

See Also

tx_event_flags_create, tx_event_flags_delete, tx_event_flags_get, tx_event_flags_
info_get, tx_event_flags_performance_system_info_get, tx_event_flags_set,
tx_event_flags_set_notify

tx_event_flags_performance_system_info_get

Retrieve performance system information

Prototype

```
UINT tx_event_flags_performance_system_info_get(ULONG *sets,
  ULONG *gets,
  ULONG *suspensions,
  ULONG *timeouts);
```

Description

This service retrieves performance information about all event flags groups in the system.

NOTE:

ThreadX library and application must be built with TX_EVENT_FLAGS_ENABLE_
PERFORMANCE_INFO defined for this service to return performance information.

Input Parameters

sets	Pointer to destination for the total number of event flags set requests performed on all groups.
gets	Pointer to destination for the total number of event flags get requests performed on all groups.
suspensions	Pointer to destination for the total number of thread event flags get suspensions on all groups.
timeouts	Pointer to destination for the total number of event flags get suspension timeouts on all groups.

NOTE:

Supplying a TX_NULL for any parameter indicates that the parameter is not required.

Return Values

TX_SUCCESS (0x00)	Successful event flags system performance get.
TX_FEATURE_NOT_ENABLED (0xFF)	The system was not compiled with performance information enabled.

Allowed From

Initialization, threads, timers, and ISRs

Example

```
ULONG sets;
ULONG gets;
ULONG suspensions;
ULONG timeouts;
...
/* Retrieve performance information on all previously created
event flag groups. */
```

```
status = tx_event_flags_performance_system_info_get(&sets, &gets,
                                                    &suspensions,
                                                    &timeouts);
/* If status is TX_SUCCESS the performance information was
successfully retrieved. */
```

See Also

tx_event_flags_create, tx_event_flags_delete, tx_event_flags_get, tx_event_flags_
info_get, tx_event_flags_performance_info_get, tx_event_flags_set, tx_event_flags_
set_notify

tx_event_flags_set

Set event flags in an event flags group

Prototype

```
UINT tx_event_flags_set(TX_EVENT_FLAGS_GROUP *group_ptr,
     ULONG flags_to_set, UINT set_option)
```

Description

This service sets or clears event flags in an event flags group, depending upon the specified set-option. All suspended threads whose event flags requests are now satisfied are resumed.

Input Parameters

group_ptr	Pointer to the previously created event flags group's Control Block
flags_to_set	Specifies the event flags to set or clear based upon the set option selected.
set_option	Specifies whether the event flags specified are ANDed or ORed into the current event flags of the group. The following are valid selections: TX_AND (0×02) TX_OR (0×00)

Selecting TX_AND specifies that the specified event flags are ANDed into the current event flags in the group. This option is often used to clear event flags in a group. Otherwise, if TX_OR is specified, the specified event flags are ORed with the current event in the group.

NOTE:

The TX_OR option forces the scheduler to review the suspension list to determine whether any threads are suspended for this event flags group.

Return Values

TX_SUCCESS[5]	(0×00)	Successful event flags set.
TX_GROUP_ERROR	(0×06)	Invalid pointer to event flags group.
TX_OPTION_ERROR	(0×08)	Invalid set-option specified.

Allowed From

Initialization, threads, timers, and ISRs

Preemption Possible

Yes

Example

```
TX_EVENT_FLAGS_GROUP my_event_group;
UINT status;
...
/* Set event flags 0, 4, and 8. */
status = tx_event_flags_set(&my_event_group, 0x111, TX_OR);
/* If status equals TX_SUCCESS, the event flags have been set and any
suspended thread whose request was satisfied has been resumed. */
```

tx_event_flags_set_notify

Notify application when event flags are set

---

[5]This value is not affected by the TX_DISABLE_ERROR_CHECKING define that is used to disable API error checking.

Prototype

```
UINT tx_event_flags_set_notify(TX_EVENT_FLAGS_GROUP *group_ptr,
                VOID (*events_set_notify)(TX_EVENT_FLAGS_GROUP *));
```

Description

This service registers a notification callback function that is called whenever one or more event flags are set in the specified event flags group. The processing of the notification callback is defined by the application.

Input Parameters

group_ptr   Pointer to previously created event flags group.
events_set_notify Pointer to application's event flags set notification function. If this value is TX_NULL, notification is disabled.

Return Values

TX_SUCCESS (0x00)   Successful registration of event flags set notification.
TX_GROUP_ERROR (0x06)   Invalid event flags group pointer.
TX_FEATURE_NOT_ENABLED   The system was compiled with notification
(0xFF)   capabilities disabled.

Allowed From

Initialization, threads, timers, and ISRs

Example

```
TX_EVENT_FLAGS_GROUP my_group;
...
/* Register the "my_event_flags_set_notify" function for monitoring
event flags set in the event flags group "my_group." */
status = tx_event_flags_set_notify(&my_group,
                                   my_event_flags_set_notify);
```

```
/* If status is TX_SUCCESS the event flags set notification function
was successfully registered. */
...
void my_event_flags_set_notify TX_EVENT_FLAGS_GROUP *group_ptr)
{
/* One or more event flags was set in this group! */
}
```

See Also

tx_event_flags_create, tx_event_flags_delete, tx_event_flags_get, tx_event_flags_info_
get, tx_event_flags_performance_info_get, tx_event_flags_performance_system_info_
get, tx_event_flags_set

Interrupt Control Service

tx_interrupt_control

Enables and disables interrupts

Prototype

```
UINT tx_interrupt_control (UINT new_posture)
```

Description

This service enables or disables interrupts as specified by the parameter *new_posture*.

NOTE:

If this service is called from an application thread, the interrupt posture remains part of that thread's context. For example, if the thread calls this routine to disable interrupts and then suspends, when it is resumed, interrupts are disabled again.

WARNING:

Do not use this service to enable interrupts during initialization! Doing so could cause unpredictable results.

Input Parameter

new_posture[1]   This parameter specifies whether interrupts are disabled or enabled. Legal values include **TX_INT_DISABLE** and **TX_INT_ENABLE**. The actual values for this parameter are port-specific. In addition, some processing architectures might support additional interrupt disable postures.

Return Values

previous posture   This service returns the previous interrupt posture to the caller. This allows users of the service to restore the previous posture after interrupts are disabled.

Allowed From

Threads, timers, and ISRs

Preemption Possible

No

Example

```
UINT my_old_posture;
...
/* Lockout interrupts */
my_old_posture = tx_interrupt_control(TX_INT_DISABLE);
/* Perform critical operations that need interrupts locked-out. */
...
/* Restore previous interrupt lockout posture. */
tx_interrupt_control(my_old_posture);
```

---

[1]This value is processor-specific and is defined in the file **tx_port.h**. This value typically maps directly to the interrupt lockout/enable bits in the processor's status register. The user must take care in selecting an appropriate value for new_posture. For the MIPS processor, TX_INT_DISABLE is 0x00 and TX_INT_ENABLE is 0x01, which corresponds to the IE bit in the Status Register (see Figure 5.2).

Mutex Services

The mutex services described in this appendix include:

tx_mutex_create	Create a mutual exclusion mutex
tx_mutex_delete	Delete a mutual exclusion mutex
tx_mutex_get	Obtain ownership of a mutex
tx_mutex_info_get	Retrieve information about a mutex
tx_mutex_performance_info_get	Get mutex performance information
tx_mutex_performance_system_info_get	Get mutex system performance information
tx_mutex_prioritize	Prioritize the mutex suspension list
tx_mutex_put	Release ownership of a mutex

tx_mutex_create

Create a mutual exclusion mutex

Prototype

```
UINT tx_mutex_create(TX_MUTEX *mutex_ptr,
     CHAR *name_ptr, UINT priority_inherit)
```

Description

This service creates a mutex for inter-thread mutual exclusion for resource protection. This service initializes the Mutex Control Block through the parameter mutex_ptr.

Input Parameters

`mutex_ptr`	Pointer to a Mutex Control Block.
`name_ptr`	Pointer to the name of the mutex.
`priority_inherit`	Specifies whether or not this mutex supports priority inheritance. If this value is TX_INHERIT, then priority inheritance is supported. However, if TX_NO_INHERIT is specified, priority inheritance is not supported by this mutex.

Return Values

TX_SUCCESS[1]	(0x00)	Successful mutex creation.
TX_MUTEX_ERROR	(0x1C)	Invalid mutex pointer. Either the pointer is NULL or the mutex has already been created.
TX_CALLER_ERROR	(0x13)	Invalid caller of this service.
TX_INHERIT_ERROR	(0x1F)	Invalid priority inheritance parameter.

Allowed From

Initialization and threads

Preemption Possible

No

Example

```
TX_MUTEX my_mutex;
UINT status;
/* Create a mutex to provide protection over a common resource. */
status = tx_mutex_create(&my_mutex, "my_mutex_name",
                    TX_NO_INHERIT);
/* If status equals TX_SUCCESS, my_mutex is ready for use. */
```

---

[1]This value is not affected by the TX_DISABLE_ERROR_CHECKING define that is used to disable API error checking.

tx_mutex_delete

Delete a mutual exclusion mutex

Prototype

```
UINT tx_mutex_delete(TX_MUTEX *mutex_ptr)
```

Description

This service deletes the specified mutex. All threads suspended waiting for the mutex are resumed and receive a TX_DELETED return status.

WARNING:

It is the application's responsibility to prevent use of a deleted mutex

---

Input Parameter

mutex_ptr   Pointer to a previously created mutex's Control Block.

Return Values

TX_SUCCESS[2]	(0x00)	Successful mutex deletion.
TX_MUTEX_ERROR	(0x1C)	Invalid mutex pointer.
TX_CALLER_ERROR	(0x13)	Invalid caller of this service.

Allowed From

Threads

Preemption Possible

Yes

---

[2]This value is not affected by the TX_DISABLE_ERROR_CHECKING define that is used to disable API error checking.

Example

```
TX_MUTEX my_mutex;
UINT status;
   ...
/* Delete a mutex. Assume that the mutex has already been created.
*/
status = tx_mutex_delete(&my_mutex);
/* If status equals TX_SUCCESS, the mutex is deleted. */
```

tx_mutex_get

Obtain ownership of a mutex

Prototype

```
UINT tx_mutex_get(TX_MUTEX *mutex_ptr, ULONG wait_option)
```

Description

This service attempts to obtain exclusive ownership of the specified mutex. If the calling thread already owns the mutex, an internal counter is incremented and a successful status is returned. If the mutex is owned by another thread and the calling thread has higher priority and priority inheritance was enabled upon mutex creation, the lower-priority thread's priority becomes temporarily raised to that of the calling thread. This service may modify the mutex Control Block through the parameter mutex_ptr.

WARNING:

Note that the priority of the lower-priority thread owning a mutex with priority-inheritance should never be modified by an external thread during mutex ownership.

Input Parameters

mutex_ptr	Pointer to a previously created mutex's Control Block.
wait_option	Defines how the service behaves if the mutex is already owned by another thread. The wait options are defined as follows:
	TX_NO_WAIT (0x00000000)

TX_WAIT_FOREVER (0xFFFFFFFF)

timeout value (0x00000001 to 0xFFFFFFFE, inclusive)

Selecting TX_NO_WAIT results in an immediate return from this service regardless of whether or not it was successful. This is the only valid option if the service is called from initialization. Selecting TX_WAIT_FOREVER causes the calling thread to suspend indefinitely until the mutex becomes available. Selecting a numeric value (1-0xFFFFFFFE) specifies the maximum number of timer-ticks to stay suspended while waiting for the mutex.

Return Values

TX_SUCCESS[3]	(0x00)	Successful mutex get operation.
TX_DELETED	(0x01)	Mutex was deleted while thread was suspended.
TX_NOT_AVAILABLE	(0x1D)	Service was unable to get ownership of the mutex.
TX_WAIT_ABORTED	(0x1A)	Suspension was aborted by another thread, timer, or ISR.
TX_MUTEX_ERROR	(0x1C)	Invalid mutex pointer.
TX_WAIT_ERROR	(0x04)	A wait option other than TX_NO_WAIT was specified on a call from a non-thread.
TX_CALLER_ERROR	(0x13)	Invalid caller of this service.

Allowed From

Initialization, threads, and timers

Preemption Possible

Yes

---

[3]This value is not affected by the TX_DISABLE_ERROR_CHECKING define that is used to disable API error checking.

Example

```
TX_MUTEX my_mutex;
UINT status;
...
/* Obtain exclusive ownership of the mutex "my_mutex". If the
mutex "my_mutex" is not available, suspend until it becomes
available. */
status = tx_mutex_get(&my_mutex, TX_WAIT_FOREVER);
```

tx_mutex_info_get

Retrieve information about a mutex

Prototype

```
UINT tx_mutex_info_get(TX_MUTEX *mutex_ptr, CHAR **name,
      ULONG *count, TX_THREAD **owner,
      TX_THREAD **first_suspended,
      ULONG *suspended_count, TX_MUTEX **next_mutex)
```

Description

This service retrieves information from the specified mutex.

Input Parameter

mutex_ptr   Pointer to a previously created Mutex Control Block.

Output Parameters

Name	Pointer to destination for the pointer to the mutex name.
Count	Pointer to destination for the ownership count of the mutex.
owner	Pointer to destination for the owning thread's pointer.
first_suspended	Pointer to destination for the pointer to the thread that is first on the suspension list of this mutex.
suspended_count	Pointer to destination for the number of threads currently suspended on this mutex.
next_mutex	Pointer to destination for the pointer of the next created mutex.

Return Values

TX_SUCCESS[4]	(0x00)	Successful mutex information retrieval.
TX_MUTEX_ERROR	(0x1C)	Invalid mutex pointer.
TX_PTR_ERROR	(0x03)	Invalid pointer (NULL) for any destination pointer.

Allowed From

Initialization, threads, timers, and ISRs

Preemption Possible

No

Example

```
TX_MUTEX my_mutex;
CHAR *name;
ULONG count;
TX_THREAD *owner;
TX_THREAD *first_suspended;
ULONG suspended_count;
TX_MUTEX *next_mutex;
UINT status;
...
/* Retrieve information about the previously created mutex "my_
mutex." */
status = tx_mutex_info_get(&my_mutex, &name,
                      &count, &owner, &first_suspended,
                      &suspended_count, &next_mutex);
/* If status equals TX_SUCCESS, the information requested is
valid. */
```

tx_mutex_performance_info_get

Get mutex performance information

---

[4]This value is not affected by the TX_DISABLE_ERROR_CHECKING define that is used to disable API error checking.

Prototype

```
UINT tx_mutex_performance_info_get(TX_MUTEX *mutex_ptr,
     ULONG *puts,

     ULONG *gets,
     ULONG *suspensions, ULONG *timeouts,
     ULONG *inversions,
     ULONG *inheritances);
```

Description

This service retrieves performance information about the specified mutex.

NOTE:

The ThreadX library and application must be built with TX_MUTEX_ENABLE_PERFORMANCE_INFO defined for this service to return performance information.

Input Parameters

mutex_ptr	Pointer to previously created mutex.
puts	Pointer to destination for the number of put requests performed on this mutex.
gets	Pointer to destination for the number of get requests performed on this mutex.
suspensions	Pointer to destination for the number of thread mutex get suspensions on this mutex.
timeouts	Pointer to destination for the number of mutex get suspension timeouts on this mutex.
inversions	Pointer to destination for the number of thread priority inversions on this mutex.
inheritances	Pointer to destination for the number of thread priority inheritance operations on this mutex.

NOTE:

Supplying a TX_NULL for any parameter indicates that the parameter is not required.

Return Values

TX_SUCCESS (0x00)	Successful mutex performance get.
TX_PTR_ERROR (0x03)	Invalid mutex pointer.
TX_FEATURE_NOT_ENABLED (0xFF)	The system was not compiled with performance information enabled.

Allowed From

Initialization, threads, timers, and ISRs

Example

```
TX_MUTEX my_mutex;
ULONG puts;
ULONG gets;
ULONG suspensions;
ULONG timeouts;
ULONG inversions;
ULONG inheritances;
...
/* Retrieve performance information on the previously created
mutex. */
status = tx_mutex_performance_info_get (&my_mutex_ptr, &puts,
     &gets,
     &suspensions,
     &timeouts, &inversions,
     &inheritances);
/* If status is TX_SUCCESS the performance information was
successfully retrieved. */
```

See Also

tx_mutex_create, tx_mutex_delete, tx_mutex_get, tx_mutex_info_get, tx_mutex_performance_system_info_get, tx_mutex_prioritize, tx_mutex_put

tx_mutex_performance_system_info_get

Get mutex system performance information

Prototype

```
UINT tx_mutex_performance_system_info_get(ULONG *puts,
      ULONG *gets,

      ULONG *suspensions,
      ULONG *timeouts,
      ULONG *inversions,
      ULONG *inheritances);
```

Description

This service retrieves performance information about all the mutexes in the system.

NOTE:

The ThreadX library and application must be built with TX_MUTEX_ENABLE_
PERFORMANCE_INFO defined for this service to return performance information.

Input Parameters

puts	Pointer to destination for the total number of put requests performed on all mutexes.
gets	Pointer to destination for the total number of get requests performed on all mutexes.
suspensions	Pointer to destination for the total number of thread mutex get suspensions on all mutexes.
timeouts	Pointer to destination for the total number of mutex get suspension timeouts on all mutexes.
inversions	Pointer to destination for the total number of thread priority inversions on all mutexes.
inheritances	Pointer to destination for the total number of thread priority inheritance operations on all mutexes.

NOTE:

Supplying a TX_NULL for any parameter indicates that the parameter is not required.

Return Values

TX_SUCCESS (0x00)   Successful mutex system performance get.

TX_FEATURE_NOT_ENABLED   The system was not compiled with
(0xFF)   performance information enabled.

Allowed From

Initialization, threads, timers, and ISRs

Example

```
ULONG puts;
ULONG gets;
ULONG suspensions;
ULONG timeouts;
ULONG inversions;
ULONG inheritances;
...
/* Retrieve performance information on all previously created
mutexes. */
status = tx_mutex_performance_system_info_get(&puts, &gets,
     &suspensions,

     &timeouts, &inversions,
     &inheritances);
/* If status is TX_SUCCESS the performance information was
successfully retrieved. */
```

See Also

tx_mutex_create, tx_mutex_delete, tx_mutex_get, tx_mutex_info_get, tx_mutex_
performance_info_get, tx_mutex_prioritize, tx_mutex_put

tx_mutex_prioritize

Prioritize the mutex suspension list

Prototype

```
UINT tx_mutex_prioritize(TX_MUTEX *mutex_ptr)
```

Description

This service places the highest-priority thread suspended for ownership of the mutex at the front of the suspension list. All other threads remain in the same FIFO order in which they were suspended.

Input Parameter

mutex_ptr   Pointer to the previously created mutex's Control Block.

Return Values

TX_SUCCESS[5]	(0x00)	Successful mutex prioritization.
TX_MUTEX_ERROR	(0x1C)	Invalid mutex pointer.

Allowed From

Initialization, threads, timers, and ISRs

Preemption Possible

No

Example

```
TX_MUTEX my_mutex;
UINT status;
. . .
/* Ensure that the highest priority thread will receive ownership
of the mutex when it becomes available. */
status = tx_mutex_prioritize(&my_mutex);
```

---

[5]This value is not affected by the TX_DISABLE_ERROR_CHECKING define that is used to disable API error checking.

```
/* If status equals TX_SUCCESS, the highest priority suspended
thread is at the front of the list. The next tx_mutex_put call
that releases ownership of the mutex will give ownership to this
thread and wake it up. */
```

tx_mutex_put

Release ownership of a mutex

Prototype

```
UINT tx_mutex_put(TX_MUTEX *mutex_ptr)
```

Description

This service decrements the ownership count of the specified mutex. If the ownership count becomes zero, the mutex becomes available to entities attempting to acquire ownership. This service modifies the Mutex Control Block through the parameter mutex_ptr.

WARNING:

If priority inheritance was selected during mutex creation, the priority of the releasing thread will revert to the priority it had when it originally obtained ownership of the mutex. Any other priority changes made to the releasing thread during ownership of the mutex may be undone.

Input Parameter

mutex_ptr   Pointer to the previously created mutex's Control Block.

Return Values[6]

TX_SUCCESS[6]	(0x00)	Successful mutex release.
TX_NOT_OWNED[6]	(0x1E)	Mutex is not owned by caller.
TX_MUTEX_ERROR	(0x1C)	Invalid pointer to mutex.
TX_CALLER_ERROR	(0x13)	Invalid caller of this service.

---

[6]This value is not affected by the TX_DISABLE_ERROR_CHECKING define that is used to disable API error checking.

Allowed From

Initialization and threads

Preemption Possible

Yes

Example

```
TX_MUTEX my_mutex;
UINT status;
...
/* Release ownership of "my_mutex." */
status = tx_mutex_put(&my_mutex);
/* If status equals TX_SUCCESS, the mutex ownership count has been decremented and if zero,
released. */
```

Message Queue Services

The message queue services described in this appendix include:

tx_queue_create	Create a message queue
tx_queue_delete	Delete a message queue
tx_queue_flush	Empty all messages in a message queue
tx_queue_front_send	Send a message to the front of a message queue
tx_queue_info_get	Retrieve information about a message queue
tx_queue_performance_info_get	Get queue performance information
tx_queue_performance_system _info_get	Get queue system performance information
tx_queue_prioritize	Prioritize a message queue suspension list
tx_queue_receive	Get a message from a message queue
tx_queue_send	Send a message to a message queue
tx_queue_send_notify	Notify application when message is sent to queue

tx_queue_create

Create a message queue

Prototype

```
UINT tx_queue_create(TX_QUEUE *queue_ptr, CHAR *name_ptr,
  UINT message_size, VOID *queue_start,
  ULONG queue_size)
```

Description

This service creates a message queue that is typically used for inter-thread communication. This service calculates the total number of messages the queue can hold from the specified message size and the total number of bytes in the queue. This service initializes the Queue Control Block through the parameter queue_ptr.

WARNING:

If the total number of bytes specified in the queue's memory area is not evenly divisible by the specified message size, the remaining bytes in the memory area are not used.

Input Parameters

queue_ptr	Pointer to a Message Queue Control Block.
name_ptr	Pointer to the name of the message queue.
message_size	Specifies the size of each message in the queue (in ULONGs). Message
	sizes range from 1 32-bit word to 16 32-bit words. Valid message size options are defined as follows:
	TX_1_ULONG (0x01)
	TX_2_ULONG (0x02)
	TX_4_ULONG (0x04)
	TX_8_ULONG (0x08)
	TX_16_ULONG (0x10)
queue_start	Starting address of the message queue.
queue_size	Total number of bytes available for the message queue.

Return Values

TX_SUCCESS[1]	(0x00)	Successful message queue creation.

---

[1]This value is not affected by the TX_DISABLE_ERROR_CHECKING define that is used to disable API error checking.

TX_QUEUE_ERROR	(0x09)	Invalid message queue pointer. Either the pointer is NULL or the queue has already been created.
TX_PTR_ERROR	(0x03)	Invalid starting address of the message queue.
TX_SIZE_ERROR	(0x05)	Specified message queue size is invalid.
TX_CALLER_ERROR	(0x13)	Invalid caller of this service.

Allowed From

Initialization and threads

Preemption Possible

No

Example

```
TX_QUEUE my_queue;
UINT status;
…
/* Create a message queue whose total size is 2000 bytes starting
at address 0x300000. Each message in this queue is defined to be 4
32-bit words long. */
status = tx_queue_create(&my_queue, "my_queue_name",
                    TX_4_ULONG, (VOID *) 0x300000, 2000);
/* If status equals TX_SUCCESS, my_queue contains room for storing
125 messages (2000 bytes/16 bytes per message). */
```

tx_queue_delete

Delete a message queue

Prototype

```
UINT tx_queue_delete (TX_QUEUE *queue_ptr)
```

Description

This service deletes the specified message queue. All threads suspended waiting for a message from this queue are resumed and receive a TX_DELETED return status.

WARNING:

It is the application's responsibility to manage the memory area associated with the queue, which is available after this service completes. In addition, the application must not use a deleted queue.

Input Parameter

queue_ptr   Pointer to a previously created message queue's Control Block.

Return Values

TX_SUCCESS[2]	(0x00)	Successful message queue deletion.
TX_QUEUE_ERROR	(0x09)	Invalid message queue pointer.
TX_CALLER_ERROR	(0x13)	Invalid caller of this service.

Allowed From

Threads

Preemption Possible

Yes

Example

```
TX_QUEUE my_queue;
UINT status;
...
/* Delete entire message queue. Assume that the queue has already
been created with a call to tx_queue_create. */
status = tx_queue_delete(&my_queue);
/* If status equals TX_SUCCESS, the message queue is deleted. */
```

---

[2]This value is not affected by the TX_DISABLE_ERROR_CHECKING define that is used to disable API error checking.

tx_queue_flush

Empty all messages in a message queue

Prototype

```
UINT tx_queue_flush (TX_QUEUE *queue_ptr)
```

Description

This service deletes all messages stored in the specified message queue. If the queue is full, messages of all suspended threads are discarded. Each suspended thread is then resumed with a return status that indicates the message send was successful. If the queue is empty, this service does nothing. This service may modify the Queue Control Block through the parameter queue_ptr.

Input Parameter

queue_ptr   Pointer to a previously created message queue's Control Block.

Return Values

TX_SUCCESS[3]	(0x00)	Successful message queue flush.
TX_QUEUE_ERROR	(0x09)	Invalid message queue pointer.
TX_CALLER_ERROR	(0x13)	Invalid caller of this service.

Allowed From

Initialization, threads, timers, and ISRs

Preemption Possible

Yes

Example

```
TX_QUEUE my_queue;
UINT status;
...
```

---

[3]This value is not affected by the TX_DISABLE_ERROR_CHECKING define that is used to disable API error checking.

```
/* Flush out all pending messages in the specified message queue.
Assume that the queue has already been created with a call to tx_
queue_create. */
status = tx_queue_flush(&my_queue);
/* If status equals TX_SUCCESS, the message queue is empty. */
```

tx_queue_front_send

Send a message to the front of a queue

Prototype

```
UINT tx_queue_front_send(TX_QUEUE *queue_ptr, VOID *source_ptr,
                         ULONG wait_option)
```

Description

This service sends a message to the front location of the specified message queue. The message is copied to the front of the queue from the memory area specified by the source pointer. This service modifies the Queue Control Block through the parameter queue_ptr.

Input Parameters

queue_ptr	Pointer to a previously created message queue's Control Block.
source_ptr	Pointer to the message.
wait_option	Defines how the service behaves if the message queue is full. The wait options are defined as follows:
	TX_NO_WAIT (0x00000000)
	TX_WAIT_FOREVER (0xFFFFFFFF)
	timeout value (0x00000001 to 0xFFFFFFFE, inclusive)
	Selecting TX_NO_WAIT results in an immediate return from this service regardless of whether or not it was successful. *This is the only valid option if the service is called from a non-thread; e.g., initialization, timer, or ISR.* Selecting TX_WAIT_FOREVER causes the calling thread to suspend indefinitely until there is room in the queue. Selecting a numeric value (1-0xFFFFFFFE) specifies the maximum number of timer-ticks to stay suspended while waiting for room in the queue.

Return Values

TX_SUCCESS[4]	(0x00)	Successful send of message.
TX_DELETED	(0x01)	Message queue was deleted while thread was suspended.
TX_QUEUE_FULL	(0x0B)	Service was unable to send message because the queue was full.
TX_WAIT_ABORTED	(0x1A)	Suspension was aborted by another thread, timer, or ISR.
TX_QUEUE_ERROR	(0x09)	Invalid message queue pointer.
TX_PTR_ERROR	(0x03)	Invalid source pointer for message.
TX_WAIT_ERROR	(0x04)	A wait option other than TX_NO_WAIT was specified on a call from a non-thread.

Allowed From

Initialization, threads, timers, and ISRs

Preemption Possible

Yes

Example

```
TX_QUEUE my_queue;
UINT status;
ULONG my_message[4];
...
/* Send a message to the front of "my_queue." Return immediately,
regardless of success. This wait option is used for calls from
initialization, timers, and ISRs. */
status = tx_queue_front_send(&my_queue, my_message, TX_NO_WAIT);
/* If status equals TX_SUCCESS, the message is at the front of the
specified queue. */
```

---

[4]This value is not affected by the TX_DISABLE_ERROR_CHECKING define that is used to disable API error checking.

tx_queue_info_get

Retrieve information about a queue

Prototype

```
UINT tx_queue_info_get(TX_QUEUE *queue_ptr, CHAR **name,
  ULONG *enqueued, ULONG *available_storage
  TX_THREAD **first_suspended,
  ULONG *suspended_count,
  TX_QUEUE **next_queue)
```

Description

This service retrieves information about the specified message queue.

Input Parameter

queue_ptr   Pointer to a previously created message queue's Control Block.

Output Parameters

name	Pointer to destination for the pointer to the queue's name.
enqueued	Pointer to destination for the number of messages currently in the queue.
available_storage	Pointer to destination for the number of messages the queue currently has space for.
first_suspended	Pointer to destination for the pointer to the thread that is first on the suspension list of this queue.
suspended_count	Pointer to destination for the number of threads currently suspended on this queue.
next_queue	Pointer to destination for the pointer of the next created queue.

Return Values

TX_SUCCESS[5]	(0x00)	Successful queue information retrieval.
TX_QUEUE_ERROR	(0x09)	Invalid message queue pointer.
TX_PTR_ERROR	(0x03)	Invalid pointer (NULL) for any destination pointer.

Allowed From

Initialization, threads, timers, and ISRs

Preemption Possible

No

Example

```
TX_QUEUE my_queue;
CHAR *name;
ULONG enqueued;
TX_THREAD *first_suspended;
ULONG suspended_count;
ULONG available_storage;
TX_QUEUE *next_queue;
UINT status;
...
/* Retrieve information about the previously created message queue
"my_queue." */
status = tx_queue_info_get(&my_queue, &name, &enqueued,
  &available_storage, &first_suspended,
  &suspended_count, &next_queue);
/* If status equals TX_SUCCESS, the information requested is valid. */
```

tx_queue_performance_info_get

Get queue performance information

---

[5]This value is not affected by the TX_DISABLE_ERROR_CHECKING define that is used to disable API error checking.

Prototype

```
UINT tx_queue_performance_info_get(TX_QUEUE *queue_ptr,
  ULONG *messages_sent,
  ULONG *messages_received,
  ULONG *empty_suspensions,
  ULONG *full_suspensions,
  ULONG *full_errors,
  ULONG *timeouts);
```

Description

This service retrieves performance information about the specified queue.

NOTE:

The ThreadX library and application must be built with TX_QUEUE_ENABLE_ PERFORMANCE_INFO defined for this service to return performance information.

Input Parameters

queue_ptr	Pointer to previously created queue.
messages_sent	Pointer to destination for the number of send requests performed on this queue.
messages_received	Pointer to destination for the number of receive requests performed on this queue.
empty_suspensions	Pointer to destination for the number of queue empty suspensions on this queue.
full_suspensions	Pointer to destination for the number of queue full suspensions on this queue.
full_errors	Pointer to destination for the number of queue full errors on this queue.
timeouts	Pointer to destination for the number of thread suspension timeouts on this queue.

NOTE:

Supplying a TX_NULL for any parameter indicates that the parameter is not required.

Return Values

TX_SUCCESS (0x00)	Successful queue performance get.
TX_PTR_ERROR (0x03)	Invalid queue pointer.
TX_FEATURE_NOT_ENABLED (0xFF)	The system was not compiled with performance information enabled.

Allowed From

Initialization, threads, timers, and ISRs

Example

```
TX_QUEUE my_queue;
ULONG messages_sent;
ULONG messages_received;
ULONG empty_suspensions;
ULONG full_suspensions;
ULONG full_errors;
ULONG timeouts;
...
/* Retrieve performance information on the previously created
queue. */
status = tx_queue_performance_info_get(&my_queue, &messages_sent,
  &messages_received,
  &empty_suspensions,
  &full_suspensions, &full_errors, &timeouts);
/* If status is TX_SUCCESS the performance information was
successfully retrieved. */
```

See Also

tx_queue_create, tx_queue_delete, tx_queue_flush, tx_queue_front_send, tx_queue_info_get, tx_queue_performance_system_info_get, tx_queue_prioritize, tx_queue_receive, tx_queue_send, tx_queue_send_notify

tx_queue_performance_system_info_get

Get queue system performance information

Prototype

```
UINT tx_queue_performance_system_info_get(ULONG *messages_sent,
  ULONG *messages_received,
  ULONG *empty_suspensions,
  ULONG *full_suspensions,
  ULONG *full_errors,
  ULONG *timeouts);
```

Description

This service retrieves performance information about all the queues in the system.

NOTE:

The ThreadX library and application must be built with TX_QUEUE_ENABLE_
PERFORMANCE_INFO defined for this service to return performance information.

Input Parameters

messages_sent	Pointer to destination for the total number of send requests performed on all queues.
messages_received	Pointer to destination for the total number of receive requests performed on all queues.
empty_suspensions	Pointer to destination for the total number of queue empty suspensions on all queues.
full_suspensions	Pointer to destination for the total number of queue full suspensions on all queues.
full_errors	Pointer to destination for the total number of queue full errors on all queues.
timeouts	Pointer to destination for the total number of thread suspension timeouts on all queues.

NOTE:

Supplying a TX_NULL for any parameter indicates that the parameter is not required.

Return Values

TX_SUCCESS (0x00)	Successful queue system performance get.
TX_FEATURE_NOT_ENABLED (0xFF)	The system was not compiled with performance information enabled.

Allowed From

Initialization, threads, timers, and ISRs

Example

```
ULONG messages_sent;
ULONG messages_received;
ULONG empty_suspensions;
ULONG full_suspensions;
ULONG full_errors;
ULONG timeouts;
...
/* Retrieve performance information on all previously created queues. */
status = tx_queue_performance_system_info_get(&messages_sent,
  &messages_received,
  &empty_suspensions,
  &full_suspensions,
  &full_errors, &timeouts);
/* If status is TX_SUCCESS the performance information was
successfully retrieved. */
```

See Also

tx_queue_create, tx_queue_delete, tx_queue_flush, tx_queue_front_send, tx_queue_info_get, tx_queue_performance_info_get, tx_queue_prioritize, tx_queue_receive, tx_queue_send, tx_queue_send_notify

tx_queue_prioritize

Prioritize the queue suspension list

Prototype

```
UINT tx_queue_prioritize(TX_QUEUE *queue_ptr)
```

Description

This service places the highest-priority thread suspended to get a message (or to place a message) on this queue at the front of the suspension list. All other threads remain in the same FIFO order in which they were suspended.

Input Parameter

queue_ptr   Pointer to a previously created message queue's Control Block.

Return Values

TX_SUCCESS[6]	(0x00)	Successful queue prioritization.
TX_QUEUE_ERROR	(0x09)	Invalid message queue pointer.

Allowed From

Initialization, threads, timers, and ISRs

Preemption Possible

No

Example

```
TX_QUEUE my_queue;
UINT status;
...
/* Ensure that the highest priority thread will receive the next
message placed on this queue. */
status = tx_queue_prioritize(&my_queue);
/* If status equals TX_SUCCESS, the highest priority suspended
thread is at the front of the list. The next tx_queue_send or
tx_queue_front_send call made to this queue will wake up this
thread. */
```

tx_queue_receive

Get a message from a message queue

---

[6]This value is not affected by the TX_DISABLE_ERROR_CHECKING define that is used to disable API error checking.

Prototype

```
UINT tx_queue_receive(TX_QUEUE *queue_ptr, VOID *destination_ptr,
                      ULONG wait_option)
```

Description

This service retrieves a message from the specified message queue. The retrieved message is copied from the queue into the memory area specified by the destination pointer. That message is then removed from the queue. This service may modify the Queue Control Block through the parameter queue_ptr.

WARNING:

The specified destination memory area must be large enough to hold the message; i.e., the message destination pointed to by destination_ptr must be at least as large as the message size for this queue. Otherwise, if the destination is not large enough, memory corruption occurs in the memory area following the destination.

Input Parameters

queue_ptr	Pointer to a previously created message queue's Control Block.
wait_option	Defines how the service behaves if the message queue is empty. The wait options are defined as follows:
	TX_NO_WAIT (0x00000000)
	TX_WAIT_FOREVER (0xFFFFFFFF)
	timeout value (0x00000001 to 0xFFFFFFFE, inclusive)
	Selecting TX_NO_WAIT results in an immediate return from this service regardless of whether or not it was successful. This is the only valid option if the service is called from a non-thread; e.g., initialization, timer, or ISR. Selecting TX_WAIT_FOREVER causes the calling thread to suspend indefinitely until a message becomes available. Selecting a numeric value (1-0xFFFFFFFE) specifies the maximum number of timer-ticks to stay suspended while waiting for a message.

Output Parameter

destination_ptr	Location of memory area to receive a copy of the message.

Return Values

TX_SUCCESS[7]	(0x00)	Successful retrieval of message.
TX_DELETED	(0x01)	Message queue was deleted while thread was suspended.
TX_QUEUE_EMPTY	(0x0A)	Service was unable to retrieve a message because the queue was empty.
TX_WAIT_ABORTED[7]	(0x1A)	Suspension was aborted by another thread, timer, or ISR.
TX_QUEUE_ERROR	(0x09)	Invalid message queue pointer.
TX_PTR_ERROR	(0x03)	Invalid destination pointer for message.
TX_WAIT_ERROR	(0x04)	A wait option other than TX_NO_WAIT was specified on a call from a non-thread.

Allowed From

Initialization, threads, timers, and ISRs

Preemption Possible

Yes

Example

```
TX_QUEUE my_queue;
UINT status;
ULONG my_message[4];
...
/* Retrieve a message from "my_queue." If the queue is empty,
suspend until a message is present. Note that this suspension is
only possible from application threads. */
status = tx_queue_receive(&my_queue, my_message,
                        TX_WAIT_FOREVER);
/* If status equals TX_SUCCESS, the message is in "my_message." */
```

---

[7]This value is not affected by the TX_DISABLE_ERROR_CHECKING define that is used to disable API error checking.

tx_queue_send

Send a message to a message queue

Prototype

```
UINT tx_queue_send(TX_QUEUE *queue_ptr, VOID *source_ptr,
                   ULONG wait_option)
```

Description

This service sends a message to the specified message queue. The sent message is copied to the queue from the memory area specified by the source pointer. This service may modify the Queue Control Block through the parameter queue_ptr.

Input Parameters

queue_ptr	Pointer to a previously created message queue's Control Block.
source_ptr	Pointer to the message.
wait_option	Defines how the service behaves if the message queue is full. The wait options are defined as follows:

TX_NO_WAIT (0x00000000)

TX_WAIT_FOREVER (0xFFFFFFFF)

timeout value (0x00000001 to 0xFFFFFFFE, inclusive)

Selecting TX_NO_WAIT results in an immediate return from this service regardless of whether or not it was successful. *This is the only valid option if the service is called from a non-thread; e.g., initialization, timer, or ISR.* Selecting TX_WAIT_FOREVER causes the calling thread to suspend indefinitely until there is room in the queue. Selecting a numeric value (1-0xFFFFFFFE) specifies the maximum number of timer-ticks to stay suspended while waiting for room in the queue.

Return Values

TX_SUCCESS[8]	(0x00)	Successful sending of message.
TX_DELETED	(0x01)	Message queue was deleted while thread was suspended.
TX_QUEUE_FULL	(0x0B)	Service was unable to send message because the queue was full.
TX_WAIT_ABORTED	(0x1A)	Suspension was aborted by another thread, timer, or ISR.
TX_QUEUE_ERROR	(0x09)	Invalid message queue pointer.
TX_PTR_ERROR	(0x03)	Invalid source pointer for message.
TX_WAIT_ERROR	(0x04)	A wait option other than TX_NO_WAIT was specified on a call from a non-thread.

Allowed From

Initialization, threads, timers, and ISRs

Preemption Possible

Yes

Example

```
TX_QUEUE my_queue;
UINT status;
ULONG my_message[4];
...
/* Send a message to "my_queue." Return immediately, regardless of
success. This wait option is used for calls from initialization,
timers, and ISRs. */
status = tx_queue_send(&my_queue, my_message, TX_NO_WAIT);
/* If status equals TX_SUCCESS, the message is in the queue. */
```

---

[8]This value is not affected by the TX_DISABLE_ERROR_CHECKING define that is used to disable API error checking.

tx_queue_send_notify

Notify application when message is sent to queue

Prototype

```
UINT tx_queue_send_notify(TX_QUEUE *queue_ptr,
  VOID (*queue_send_notify)(TX_QUEUE *));
```

Description

This service registers a notification callback function that is called whenever a message is sent to the specified queue. The processing of the notification callback is defined by the application.

Input Parameters

queue_ptr   Pointer to previously created queue.

queue_send_notify   Pointer to application's queue send notification function. If this value is TX_NULL, notification is disabled.

Return Values

TX_SUCCESS (0x00)   Successful registration of queue send notification.

TX_QUEUE_ERROR (0x09)   Invalid queue pointer.

TX_FEATURE_NOT_ENABLED (0xFF)   The system was compiled with notification capabilities disabled.

Allowed From

Initialization, threads, timers, and ISRs

Example

```
TX_QUEUE my_queue;
...
/* Register the "my_queue_send_notify" function for monitoring
messages sent to the queue "my_queue." */
```

```
status = tx_queue_send_notify(&my_queue, my_queue_send_notify);
/* If status is TX_SUCCESS the queue send notification function was
successfully registered. */
…
void my_queue_send_notify(TX_QUEUE *queue_ptr)
{
/* A message was just sent to this queue! */
}
```

See Also

tx_queue_create, tx_queue_delete, tx_queue_flush, tx_queue_front_send, tx_queue_
info_get, tx_queue_performance_info_get, tx_queue_performance_system_info_get,
tx_queue_prioritize, tx_queue_receive, tx_queue_send

Counting Semaphore Services

The counting semaphore services described in this appendix are:

tx_semaphore_ceiling_put	Place an instance in counting semaphore with ceiling
tx_semaphore_create	Create a counting semaphore
tx_semaphore_delete	Delete a counting semaphore
tx_semaphore_get	Get an instance from a counting semaphore
tx_semaphore_info_get	Retrieve information about a counting semaphore
tx_semaphore_performance_info_get	Get semaphore performance information
tx_semaphore_performance_system_info_get	Get semaphore system performance information
tx_semaphore_prioritize	Prioritize a counting semaphore suspension list
tx_semaphore_put	Place an instance in a counting semaphore
tx_semaphore_put_notify	Notify application when semaphore is put

tx_semaphore_ceiling_put

Place an instance in counting semaphore with ceiling

Prototype

```
UINT tx_semaphore_ceiling_put(TX_SEMAPHORE *semaphore_ptr,
                              ULONG ceiling);
```

Description

This service puts an instance into the specified counting semaphore, which in reality increments the counting semaphore by one. If the counting semaphore's current value is greater than or equal to the specified ceiling, the instance will not be put and a TX_CEILING_EXCEEDED error will be returned.

Input Parameters

semaphore_ptr   Pointer to previously created semaphore.

ceiling   Maximum limit allowed for the semaphore (valid values range from 1 through 0xFFFFFFFF).

Return Values

TX_SUCCESS	(0x00)	Successful semaphore ceiling put.
TX_CEILING_EXCEEDED	(0x21)	Put request exceeds ceiling.
TX_INVALID_CEILING	(0x22)	An invalid value of zero was supplied for ceiling.
TX_SEMAPHORE_ERROR	(0x03)	Invalid semaphore pointer.

Allowed From

Initialization, threads, timers, and ISRs

Example

```
TX_SEMAPHORE my_semaphore;
...
/* Increment the counting semaphore "my_semaphore" but make sure
that it never exceeds 7 as specified in the call. */
status = tx_semaphore_ceiling_put(&my_semaphore, 7);
```

```
/* If status is TX_SUCCESS the semaphore count has been
incremented. */
```

See Also

tx_semaphore_create, tx_semaphore_delete, tx_semaphore_get, tx_semaphore_info_get, tx_semaphore_performance_info_get, tx_semaphore_performance_system_info_get, tx_semaphore_prioritize, tx_semaphore_put, tx_semaphore_put_notify

tx_semaphore_create

Create a counting semaphore

Prototype

```
UINT tx_semaphore_create(TX_SEMAPHORE *semaphore_ptr,
                         CHAR *name_ptr, ULONG initial_count)
```

Description

This service creates a counting semaphore for inter-thread synchronization. The initial semaphore count is specified as a parameter. This service initializes the Semaphore Control Block through the parameter semaphore_ptr.

Input Parameters

emaphore_ptr	Pointer to a Semaphore Control Block.
name_ptr	Pointer to the name of the semaphore.
initial_count	Specifies the initial count for this semaphore. Legal values are from 0x00000000 to 0xFFFFFFFF (inclusive).

Return Values

TX_SUCCESS[1]	(0x00)	Successful semaphore creation

---

[1]This value is not affected by the TX_DISABLE_ERROR_CHECKING define that is used to disable API error checking.

TX_SEMAPHORE_ERROR (0x0C)   Invalid semaphore pointer. Either the pointer is NULL or the semaphore has already been created.

TX_CALLER_ERROR   (0x13)   Invalid caller of this service.

Allowed From

Initialization and threads

Preemption Possible

No

Example

```
TX_SEMAPHORE my_semaphore;
UINT status;
...
/* Create a counting semaphore with an initial value of 1. This is
typically the technique used to create a binary semaphore. Binary
semaphores are used to provide protection over a common resource. */
status = tx_semaphore_create(&my_semaphore, "my_semaphore_name", 1);
/* If status equals TX_SUCCESS, my_semaphore is ready for use. */
```

tx_semaphore_delete

Delete a counting semaphore

Prototype

```
UINT tx_semaphore_delete (TX_SEMAPHORE *semaphore_ptr)
```

Description

This service deletes the specified counting semaphore. All threads suspended waiting for an instance of this semaphore are resumed and receive a TX_DELETED return status.

WARNING:

It is the application's responsibility to prevent use of a deleted semaphore.

Parameter

semaphore_ptr   Pointer to a previously created semaphore's Control Block.

Return Values

TX_SUCCESS[2]   (0x00)   Successful counting semaphore deletion.
TX_SEMAPHORE_ERROR   (0x0C)   Invalid counting semaphore pointer.
TX_CALLER_ERROR   (0x13)   Invalid caller of this service.

Allowed From

Threads

Preemption Possible

Yes

Example

```
TX_SEMAPHORE my_semaphore;
UINT status;
...
/* Delete counting semaphore. Assume that the counting semaphore
has already been created. */
status = tx_semaphore_delete(&my_semaphore);
/* If status equals TX_SUCCESS, the counting semaphore is deleted. */
```

tx_semaphore_get

Get an instance from a counting semaphore

Prototype

```
UINT tx_semaphore_get(TX_SEMAPHORE *semaphore_ptr,
                      ULONG wait_option)
```

---

[2]This value is not affected by the TX_DISABLE_ERROR_CHECKING define that is used to disable API error checking.

Description

This service retrieves an instance (a single count) from the specified counting semaphore. As a result, the specified semaphore's count is decreased by one. This service may modify the semaphore's Control Block through the parameter semaphore_ptr.

Input Parameters

semaphore_ptr   Pointer to a previously created counting semaphore's Control Block.

wait_option   Defines how the service behaves if there are no instances of the semaphore available; i.e., the semaphore count is zero. The wait options are defined as follows:

TX_NO_WAIT (0x00000000)

TX_WAIT_FOREVER (0xFFFFFFFF)

timeout value (0x00000001 to 0xFFFFFFFE, inclusive)

Selecting TX_NO_WAIT results in an immediate return from this service regardless of whether or not it was successful. This is the only valid option if the service is called from a non-thread; e.g., initialization, timer, or ISR.

Selecting TX_WAIT_FOREVER causes the calling thread to suspend indefinitely until a semaphore instance becomes available. Selecting a numeric value (1-0xFFFFFFFE) specifies the maximum number of timer-ticks to stay suspended while waiting for a semaphore instance.

Return Values

TX_SUCCESS [3]	(0x00)	Successful retrieval of a semaphore instance.
TX_DELETED	(0x01)	Counting semaphore was deleted while thread was suspended.
TX_NO_INSTANCE	(0x0D)	Service was unable to retrieve an instance of the counting semaphore (semaphore count is zero).

---

[3]This value is not affected by the TX_DISABLE_ERROR_CHECKING define that is used to disable API error checking.

TX_WAIT_ABORTED	(0x1A)	Suspension was aborted by another thread, timer, or ISR.
TX_SEMAPHORE_ERROR	(0x0C)	Invalid counting semaphore pointer.
TX_WAIT_ERROR	(0x04)	A wait option other than TX_NO_WAIT was specified on a call from a non-thread.

Allowed From

Initialization, threads, timers, and ISRs

Preemption Possible

Yes

Example

```
TX_SEMAPHORE my_semaphore;
UINT status;
...
/* Get a semaphore instance from the semaphore "my_semaphore." If
the semaphore count is zero, suspend until an instance becomes
available. Note that this suspension is only possible from
application threads. */
status = tx_semaphore_get(&my_semaphore, TX_WAIT_FOREVER);
/* If status equals TX_SUCCESS, the thread has obtained an
instance of the semaphore. */
```

tx_semaphore_info_get

Retrieve information about a counting semaphore

Prototype

```
UINT tx_semaphore_info_get(TX_SEMAPHORE *semaphore_ptr,
                           CHAR **name, ULONG *current_value,
                           TX_THREAD **first_suspended,
                           ULONG *suspended_count,
                           TX_SEMAPHORE **next_semaphore)
```

Description

This service retrieves information about the specified semaphore.

Input Parameter

semaphore_ptr   Pointer to a previously created semaphore's Control Block.

Output Parameters

name	Pointer to destination for the pointer to the semaphore's name.
current_value	Pointer to destination for the current semaphore's count.
first_suspended	Pointer to destination for the pointer to the thread that is first on the suspension list of this semaphore.
suspended_count	Pointer to destination for the number of threads currently suspended on this semaphore.
next_semaphore	Pointer to destination for the pointer of the next created semaphore.

Return Values

TX_SUCCESS[4]	(0x00)	Successful semaphore information retrieval.
TX_SEMAPHORE_ERROR	(0x0C)	Invalid semaphore pointer.
TX_PTR_ERROR	(0x03)	Invalid pointer (NULL) for any destination pointer.

Allowed From

Initialization, threads, timers, and ISRs

Preemption Possible

No

---

[4]This value is not affected by the TX_DISABLE_ERROR_CHECKING define that is used to disable API error checking.

Example

```
TX_SEMAPHORE my_semaphore;
CHAR *name;
ULONG current_value;
TX_THREAD *first_suspended;
ULONG suspended_count;
TX_SEMAPHORE *next_semaphore;
UINT status;
...
/* Retrieve information about the previously created semaphore
"my_semaphore." */
status = tx_semaphore_info_get(&my_semaphore, &name, &current_value,
                               &first_suspended, &suspended_count,
                               &next_semaphore);
/* If status equals TX_SUCCESS, the information requested is
valid. */
```

tx_semaphore_performance_info_get

Get semaphore performance information

Prototype

```
UINT tx_semaphore_performance_info_get(TX_SEMAPHORE *semaphore_ptr,
    ULONG *puts, ULONG *gets, ULONG *suspensions, ULONG *timeouts);
```

Description

This service retrieves performance information about the specified semaphore.

NOTE:

The ThreadX library and application must be built with TX_SEMAPHORE_ENABLE_
PERFORMANCE_INFO defined for this service to return performance information.

Input Parameters

semaphore_ptr   Pointer to previously created semaphore.

puts   Pointer to destination for the number of put requests performed on this semaphore.

gets   Pointer to destination for the number of get requests performed on this semaphore.

suspensions   Pointer to destination for the number of thread suspensions on this semaphore.

timeouts   Pointer to destination for the number of thread suspension timeouts on this semaphore.

NOTE:

Supplying a TX_NULL for any parameter indicates that the parameter is not required.

Return Values

TX_SUCCESS	(0x00)	Successful semaphore performance get.
TX_PTR_ERROR	(0x03)	Invalid semaphore pointer.
TX_FEATURE_NOT_ENABLED	(0xFF)	The system was not compiled with performance information enabled.

Allowed From

Initialization, threads, timers, and ISRs

Example

```
TX_SEMAPHORE my_semaphore;
ULONG puts;
ULONG gets;
ULONG suspensions;
ULONG timeouts;
/* Retrieve performance information on the previously created
semaphore. */
status = tx_semaphore_performance_info_get(&my_semaphore, &puts,
                                    &gets, &suspensions,
                                    &timeouts);
/* If status is TX_SUCCESS the performance information was
successfully retrieved. */
```

See Also

tx_seamphore_ceiling_put, tx_semaphore_create, tx_semaphore_delete, tx_semaphore_ get, tx_semaphore_info_get, tx_semaphore_performance_system_info_get, tx_ semaphore_prioritize, tx_semaphore_put, tx_semaphore_put_notify

tx_semaphore_performance_system_info_get

Get semaphore system performance information

Prototype

```
UINT tx_semaphore_performance_system_info_get(ULONG *puts,
                                              ULONG *gets,
                                              ULONG *suspensions,
                                              ULONG *timeouts);
```

Description

This service retrieves performance information about all the semaphores in the system.

NOTE:

The ThreadX library and application must be built with TX_SEMAPHORE_ENABLE_ PERFORMANCE_INFO defined for this service to return performance information

---

Input Parameters

puts	Pointer to destination for the total number of put requests performed on all semaphores.
gets	Pointer to destination for the total number of get requests performed on all semaphores.
suspensions	Pointer to destination for the total number of thread suspensions on all semaphores.
timeouts	Pointer to destination for the total number of thread suspension timeouts on all semaphores.

Return Values

TX_SUCCESS (0x00)   Successful semaphore system performance get.
TX_FEATURE_NOT_ENABLED   The system was not compiled with performance
(0xFF)   information enabled.

Allowed From

Initialization, threads, timers, and ISRs

Example

```
ULONG puts;
ULONG gets;
ULONG suspensions;
ULONG timeouts;
...
/* Retrieve performance information on all previously created
semaphores. */
status = tx_semaphore_performance_system_info_get(&puts, &gets,
                                                &suspensions,
                                                &timeouts);
/* If status is TX_SUCCESS the performance information was
successfully retrieved. */
```

See Also

tx_seamphore_ceiling_put, tx_semaphore_create, tx_semaphore_delete, tx_semaphore_
get, tx_semaphore_info_get, tx_semaphore_performance_info_get, tx_semaphore_
prioritize, tx_semaphore_put, tx_semaphore_put_notify

tx_semaphore_prioritize

Prioritize the semaphore suspension list

Prototype

```
UINT tx_semaphore_prioritize(TX_SEMAPHORE *semaphore_ptr)
```

Description

This service places the highest-priority thread suspended for an instance of the semaphore at the front of the suspension list. All other threads remain in the same FIFO order in which they were suspended.

Input Parameter

semaphore_ptr   Pointer to a previously created semaphore's Control Block.

Return Values

TX_SUCCESS[5]   (0x00)   Successful semaphore prioritize.
TX_SEMAPHORE_ERROR   (0x0C)   Invalid counting semaphore pointer.

Allowed From

Initialization, threads, timers, and ISRs

Preemption Possible

No

Example

```
TX_SEMAPHORE my_semaphore;
UINT status;
...
/* Ensure that the highest priority thread will receive the next
instance of this semaphore. */
status = tx_semaphore_prioritize(&my_semaphore);
/* If status equals TX_SUCCESS, the highest priority suspended
thread is at the front of the list. The next tx_semaphore_put call
made to this queue will wake up this thread. */
```

tx_semaphore_put

Place an instance in a counting semaphore

---

[5]This value is not affected by the TX_DISABLE_ERROR_CHECKING define that is used to disable API error checking.

Prototype

```
UINT tx_semaphore_put(TX_SEMAPHORE *semaphore_ptr)
```

Description

Conceptually, this service puts an instance into the specified counting semaphore. In reality, this service increments the counting semaphore by one. This service modifies the Semaphore Control Block through the parameter semaphore_ptr.

WARNING:

If this service is called when the semaphore is all ones (0xFFFFFFFF), the new put operation will cause the semaphore to be reset to zero.

Input Parameter

semaphore_ptr   Pointer to the previously created counting semaphore's Control Block.

Return Values

TX_SUCCESS[6]	(0x00)	Successful semaphore put.
TX_SEMAPHORE_ERROR	(0x0C)	Invalid pointer to counting semaphore.

Allowed From

Initialization, threads, timers, and ISRs

Preemption Possible

Yes

Example

```
TX_SEMAPHORE my_semaphore;
UINT status;
...
```

---

[6]This value is not affected by the TX_DISABLE_ERROR_CHECKING define that is used to disable API error checking.

```
/* Increment the counting semaphore "my_semaphore." */
status = tx_semaphore_put(&my_semaphore);
/* If status equals TX_SUCCESS, the semaphore count has been
incremented. Of course, if a thread was waiting, it was given the
semaphore instance and resumed. */
```

tx_semaphore_put_notify

Notify application when semaphore is put

Prototype

```
UINT tx_semaphore_put_notify(TX_SEMAPHORE *semaphore_ptr,
          VOID (*semaphore_put_notify)(TX_SEMAPHORE *));
```

Description

This service registers a notification callback function that is called whenever the specified semaphore is put. The processing of the notification callback is defined by the application.

Input Parameters

semaphore_ptr   Pointer to previously created semaphore.
semaphore_put_notify   Pointer to application's semaphore put notification function. If this value is TX_NULL, notification is disabled.

Return Values

TX_SUCCESS   (0x00) Successful registration of semaphore put notification.

TX_SEMAPHORE_ERROR   (0x0C) Invalid semaphore pointer.
TX_FEATURE_NOT_ENABLED (0xFF) The system was compiled with notification capabilities disabled.

Allowed From

Initialization, threads, timers, and ISRs

Example

```
TX_SEMAPHORE my_semaphore;
...
/* Register the "my_semaphore_put_notify" function for monitoring
the put operations on the semaphore "my_semaphore." */
status = tx_semaphore_put_notify(&my_semaphore,
my_semaphore_put_notify);
/* If status is TX_SUCCESS the semaphore put notification function
was successfully registered. */
void my_semaphore_put_notify(TX_SEMAPHORE *semaphore_ptr)
{
/* The semaphore was just put! */
}
```

See Also

tx_seamphore_ceiling_put, tx_semaphore_create, tx_semaphore_delete, tx_semaphore_get, tx_semaphore_info_get, tx_semaphore_performance_info_get, tx_semaphore_performance_system_info_get, tx_semaphore_prioritize, tx_semaphore_put

Thread Services

The thread services described in this appendix include:

tx_thread_create	Create an application thread
tx_thread_delete	Delete an application thread
tx_thread_entry_exit_notify	Notify application upon thread entry and exit
tx_thread_identify	Retrieve pointer to currently executing thread
tx_thread_info_get	Retrieve information about a thread
tx_thread_performance_info_get	Get thread performance information
tx_thread_performance_system_info_get	Get thread system performance information
tx_thread_preemption_change	Change preemption-threshold of thread
tx_thread_priority_change	Change priority of an application thread
tx_thread_relinquish	Relinquish control to other application threads
tx_thread_reset	Reset thread
tx_thread_resume	Resume suspended application thread
tx_thread_sleep	Suspend current thread for specified time
tx_thread_stack_error_notify	Register thread stack error notification callback

tx_thread_suspend   Suspend an application thread

tx_thread_terminate   Terminate an application thread

tx_thread_time_slice_change   Change time-slice of application thread

tx_thread_wait_abort   Abort suspension of specified thread

tx_thread_create

Create an application thread

Prototype

```
UINT tx_thread_create(TX_THREAD *thread_ptr, CHAR *name_ptr,
                VOID (*entry_function)(ULONG),
                ULONG entry_input, VOID *stack_start,
                ULONG stack_size, UINT priority,
                UINT preempt_threshold, ULONG time_slice,
                UINT auto_start)
```

Description

This service creates an application thread, which will start execution at the specified task entry function. The stack, priority, preemption-threshold, and time-slice are among the attributes specified by the parameters. In addition, the auto_start parameter determines whether the thread starts executing immediately or is created in a suspended state. This service initializes the Thread Control Block through the parameter thread_ptr.

Input Parameters

thread_ptr	Pointer to a Thread Control Block.
name_ptr	Pointer to the name of the thread.
entry_function	Specifies the initial C function for thread execution. When a thread returns from this entry function, it is placed in a completed state and suspended indefinitely.

entry_input	A user-specified value to be passed as an argument to the thread entry function. The use of this value within the entry function is at the discretion of the user.
stack_start	Starting address of the stack's memory area.
stack_size	Number of bytes in the stack memory area. The thread's stack area must be large enough to handle its worst-case function call nesting and local variable usage.
priority[1]	Numerical priority of thread. Legal values are from 0 to TX_MAX_PRIORITIES-1 (inclusive), where a value of 0 represents the highest priority.
preempt_threshold	Highest priority level (0-31) of disabled preemption. Only priorities higher than this level are allowed to preempt this thread. This value must be less than or equal to the specified priority. A value equal to the thread priority disables preemption-threshold.
time_slice	Number of timer-ticks this thread is allowed to run before other ready threads of the same priority get a chance to run. Note that using preemption-threshold disables time-slicing. Legal time-slice values range from 1 to 0xFFFFFFFF (inclusive). A value of TX_NO_TIME_SLICE (a value of 0) disables time-slicing of this thread. Note: Using time-slicing results in a slight amount of system overhead. Since time-slicing is useful only in cases where multiple threads share the same priority, threads having a unique priority should not be assigned a time-slice.
auto_start	Specifies whether the thread starts immediately or is placed in a suspended state. Legal options are TX_AUTO_START (0x01) and TX_DONT_START (0x00). If TX_DONT_START is specified, the application must later call tx_thread_resume in order for the thread to run.

---

[1]TX_MAX_PRIORITIES is defined to be the number of priority levels. Legal values range from 32 through 1024 (inclusive) and must be evenly divisible by 32. By default, this value is set to 32 priority levels, which we will use in this book. Thus, we will assume priority values in the range from 0 to 31 (inclusive).

Return Values

TX_SUCCESS[2]	(0x00)	Successful thread creation.
TX_THREAD_ERROR	(0x0E)	Invalid thread control pointer. Either the pointer is NULL or the thread has already been created.
TX_PTR_ERROR	(0x03)	Invalid starting address of the entry point, or the stack area pointer is invalid, usually NULL.
TX_SIZE_ERROR	(0x05)	Size of stack area is invalid. Threads must have at least TX_MINIMUM_STACK[3] bytes to execute.
TX_PRIORITY_ERROR	(0x0F)	Invalid thread priority, which is a value outside the range of 0-31.
TX_THRESH_ERROR	(0x18)	Invalid preemption-threshold specified. This value must be a valid priority less than or equal to the initial priority of the thread.
TX_START_ERROR	(0x10)	Invalid auto-start selection.
TX_CALLER_ERROR	(0x13)	Invalid caller of this service.

Allowed From

Initialization and threads

Preemption Possible

Yes

Example

```
TX_THREAD my_thread;
UINT status;
...
/* Create a thread of priority 15 whose entry point is "my_thread_
entry". This bytes in size, starting at address 0x400000. The
preemption-threshold is setup to allow preemption at priorities
```

---

[2]This value is not affected by the TX_DISABLE_ERROR_CHECKING define that is used to disable API error checking.

[3]This value depends on the specific processor used and can be found in the file tx_port.h. A typical value is 200, which is too small for most applications.

```
above 10. Time-slicing is disabled. This thread is automatically
put into a ready condition. */
status = tx_thread_create(&my_thread, "my_thread",
          my_thread_entry, 0x1234, (VOID *) 0x400000, 1000,15, 10,
          TX_NO_TIME_SLICE, TX_AUTO_START);
/* If status equals TX_SUCCESS, my_thread is ready execution! */
...
/* Thread's entry function. When "my_thread" actually begins
execution, control is transferred to this function. */
VOID my_thread_entry (ULONG initial_input)
{
/* When we get here, the value of initial_input is 0x1234. See how
this was specified during creation.
The real work of the thread, including calls to other functions
should be called from here!
When/if this function returns, the corresponding thread is placed
into a "completed" state and suspended. */
}
```

tx_thread_delete

Delete an application thread

Prototype

```
UINT tx_thread_delete(TX_THREAD *thread_ptr)
```

Description

This service deletes the specified application thread. Since the specified thread must be in a terminated or completed state, this service cannot be called from a thread attempting to delete itself.

NOTE:

It is the application's responsibility to manage the memory area associated with the thread's stack, which is available after this service completes. In addition, the application must not use a deleted thread.

Input Parameter

thread_ptr   Pointer to a previously created application thread's Control Block.

Return Values

TX_SUCCESS[4] (0x00)	Successful thread deletion.
TX_THREAD_ERROR (0x0E)	Invalid application thread pointer.
TX_DELETE_ERROR3 (0x11)	Specified thread is not in a terminated or completed state.
TX_CALLER_ERROR (0x13)	Invalid caller of this service.

Allowed From

Threads and timer

Preemption Possible

No

Example

```
TX_THREAD my_thread;
  UINT status;
  ...
  /* Delete an application thread whose control block is "my_
  thread". Assume that the thread has already been created with a
  call to tx_thread_create. */
  status = tx_thread_delete(&my_thread);
  /* If status equals TX_SUCCESS, the application thread is
  deleted. */
```

tx_thread_entry_exit_notify

Notify application upon thread entry and exit

---

[4]This value is not affected by the TX_DISABLE_ERROR_CHECKING define that is used to disable API error checking.

Prototype

```
UINT tx_thread_entry_exit_notify(TX_THREAD *thread_ptr,
                  VOID (*entry_exit_notify)(TX_THREAD *, UINT))
```

Description

This service registers a notification callback function that is called whenever the specified thread is entered or exits. The processing of the notification callback is defined by the application.

Input Parameters

thread_ptr   Pointer to previously created thread.

entry_exit_notify   Pointer to application's thread entry/exit notification function. The second parameter to the entry/exit notification function designates if an entry or exit is present. The value TX_THREAD_ ENTRY (0x00) indicates the thread was entered, while the value TX_THREAD_EXIT (0x01) indicates the thread was exited. If this value is TX_NULL, notification is disabled.

Return Values

TX_SUCCESS	(0x00)	Successful registration of the thread entry/exit notification function.
TX_THREAD_ERROR	(0x0E)	Invalid thread pointer.
TX_FEATURE_NOT_ENABLED	(0xFF)	The system was compiled with notification capabilities disabled.

Allowed From

Initialization, threads, timers, and ISRs

Example

```
TX_THREAD my_thread;
/* Register the "my_entry_exit_notify" function for monitoring the
entry/exit of the thread "my_thread." */
```

```
status = tx_thread_entry_exit_notify(&my_thread,
                                    my_entry_exit_notify);
/* If status is TX_SUCCESS the entry/exit notification function
   was successfully registered. */
...
void my_entry_exit_notify(TX_THREAD *thread_ptr, UINT condition)
{
/* Determine if the thread was entered or exited. */
if (condition == TX_THREAD_ENTRY)
/* Thread entry! */
else if (condition == TX_THREAD_EXIT)
Thread exit! */
}
```

See Also

tx_thread_create, tx_thread_delete, tx_thread_entry_exit_notify, tx_thread_identify, tx_thread_info_get, tx_thread_performance_info_get, tx_thread_performance_system_info_get, tx_thread_preemption_change, tx_thread_priority_change, tx_thread_relinquish, tx_thread_reset, tx_thread_resume, tx_thread_sleep, tx_thread_stack_error_notify, tx_thread_suspend, tx_thread_terminate, tx_thread_time_slice_change, tx_thread_wait_abort

tx_thread_identify

Retrieves pointer to currently executing thread

Prototype

TX_THREAD* tx_thread_identify (VOID)

Description

This service returns a pointer to the currently executing thread's Control Block. If no thread is executing, this service returns a null pointer.

> **NOTE:**
> If this service is called from an ISR, the return value represents the thread running prior to the executing interrupt handler.

Parameters

None

Return Values

thread pointer   Pointer to the currently executing thread's Control Block. If no thread is executing, the return value is TX_NULL.

Allowed From

Threads and ISRs

Preemption Possible

No

Example

```
TX_THREAD *my_thread_ptr;
...
/* Find out who we are! */
my_thread_ptr = tx_thread_identify();
/* If my_thread_ptr is non-null, we are currently executing from
that thread or an ISR that interrupted that thread. Otherwise,
this service was called from an ISR when no thread was running
when the interrupt occurred. */
```

tx_thread_info_get

Retrieve information about a thread

Prototype

```
UINT tx_thread_info_get(TX_THREAD *thread_ptr, CHAR **name,
                        UINT *state, ULONG *run_count, UINT *priority,
```

```
UINT *preemption_threshold,
ULONG *time_slice, TX_THREAD **next_thread,
TX_THREAD **suspended_thread)
```

Description

This service retrieves information about the specified thread.

Input Parameter

thread_ptr   Pointer to a previously created thread's Control Block.

Output Parameters

name	Pointer to destination for the pointer to the thread's name.
state	Pointer to destination for the thread's current execution state. Possible values are as follows: TX_READY (0x00) TX_COMPLETED (0x01) TX_TERMINATED (0x02) TX_SUSPENDED (0x03) TX_SLEEP (0x04) TX_QUEUE_SUSP (0x05) TX_SEMAPHORE_SUSP (0x06) TX_EVENT_FLAG (0x07) TX_BLOCK_MEMORY (0x08) TX_BYTE_MEMORY (0x09) TX_MUTEX_SUSP (0x0D) TX_IO_DRIVER (0x0A)
run_count	Pointer to destination for the thread's run count.
priority	Pointer to destination for the thread's priority.
preemption_threshold	Pointer to destination for the thread's preemption-threshold.
time_slice	Pointer to destination for the thread's time-slice.
next_thread	Pointer to destination for next created thread pointer.
suspended_thread	Pointer to destination for pointer to next thread in suspension list.

Return Values

TX_SUCCESS[5]	(0x00)	Successful thread information retrieval.
TX_THREAD_ERROR	(0x0E)	Invalid thread control pointer.
TX_PTR_ERROR	(0x03)	Invalid pointer (NULL) for any destination pointer.

Allowed From

Initialization, threads, timers, and ISRs

Preemption Possible

No

Example

```
TX_THREAD my_thread;
CHAR *name;
UINT state;
ULONG run_count;
UINT priority;
UINT preemption_threshold;
UINT time_slice;
TX_THREAD *next_thread;
TX_THREAD *suspended_thread;
UINT status;
...
/* Retrieve information about the previously created thread "my_
thread." */
status = tx_thread_info_get(&my_thread, &name, &state,
                        &run_count, &priority,
                        &preemption_threshold, &time_slice,
                        &next_thread,&suspended_thread);
/* If status equals TX_SUCCESS, the information requested is
valid. */
```

---

[5]This value is not affected by the TX_DISABLE_ERROR_CHECKING define that is used to disable API error checking.

tx_thread_performance_info_get

Get thread performance information

Prototype

```
UINT tx_thread_performance_info_get(TX_THREAD *thread_ptr,
                        ULONG *resumptions,
                        ULONG *suspensions,
                        ULONG *solicited_preemptions,
                        ULONG *interrupt_preemptions,
                        ULONG *priority_inversions,
                        ULONG *time_slices,
                        ULONG *relinquishes,
                        ULONG *timeouts, ULONG *wait_aborts,
                        TX_THREAD **last_preempted_by);
```

Description

This service retrieves performance information about the specified thread.

NOTE:

The ThreadX library and application must be built with TX_THREAD_ENABLE_PERFORMANCE_INFO defined in order for this service to return performance information.

Input Parameters

thread_ptr	Pointer to previously created thread.
resumptions	Pointer to destination for the number of resumptions of this thread.
suspensions	Pointer to destination for the number of suspensions of this thread.
solicited_preemptions	Pointer to destination for the number of preemptions as a result of a ThreadX API service call made by this thread.
interrupt_preemptions	Pointer to destination for the number of preemptions of this thread as a result of interrupt processing.

priority_inversions | Pointer to destination for the number of priority inversions of this thread.

time_slices | Pointer to destination for the number of time-slices of this thread.

relinquishes | Pointer to destination for the number of thread relinquishes performed by this thread.

timeouts | Pointer to destination for the number of suspension timeouts on this thread.

wait_aborts | Pointer to destination for the number of wait aborts performed on this thread.

last_preempted_by | Pointer to destination for the thread pointer that last preempted this thread.

NOTE:

Supplying a TX_NULL for any parameter indicates that the parameter is not required.

Return Values

TX_SUCCESS	(0x00)	Successful thread performance get.
TX_PTR_ERROR	(0x03)	Invalid thread pointer.
TX_FEATURE_NOT_ENABLED	(0xFF)	The system was not compiled with performance information enabled.

Allowed From

Initialization, threads, timers, and ISRs

Example

```
TX_THREAD my_thread;
ULONG resumptions;
ULONG suspensions;
ULONG solicited_preemptions;
ULONG interrupt_preemptions;
ULONG priority_inversions;
ULONG time_slices;
ULONG relinquishes;
```

```
ULONG timeouts;
ULONG wait_aborts;
TX_THREAD *last_preempted_by;
...
/* Retrieve performance information on the previously created
thread. */
status = tx_thread_performance_info_get(&my_thread, &resumptions,
                        &suspensions,
                        &solicited_preemptions,
                        &interrupt_preemptions,
                        &priority_inversions, &time_slices,
                        &relinquishes, &timeouts,
                        &wait_aborts, &last_preempted_by);
/* If status is TX_SUCCESS the performance information was
successfully retrieved. */
```

See Also

tx_thread_create, tx_thread_delete, tx_thread_entry_exit_notify, tx_thread_identify, tx_thread_info_get, tx_thread_performance_system_info_get, tx_thread_preemption_ change, tx_thread_priority_change, tx_thread_relinquish, tx_thread_reset, tx_ thread_resume, tx_thread_sleep, tx_thread_stack_error_notify, tx_thread_suspend, tx_thread_terminate, tx_thread_time_slice_change, tx_thread_wait_abort

tx_thread_performance_system_info_get

Get thread system performance information

Prototype

```
UINT tx_thread_performance_system_info_get(ULONG *resumptions,
                        ULONG *suspensions,
                        ULONG *solicited_preemptions,
                        ULONG *interrupt_preemptions,
                        ULONG *priority_inversions,
                        ULONG *time_slices,
                        ULONG *relinquishes,
```

```
                              ULONG *timeouts,
                              ULONG *wait_aborts,
                              ULONG *non_idle_returns,
                              ULONG *idle_returns);
```

Description

This service retrieves performance information about all the threads in the system.

NOTE:

The ThreadX library and application must be built with TX_THREAD_ENABLE_ PERFORMANCE_INFO defined in order for this service to return performance information.

Input Parameters

resumptions	Pointer to destination for the total number of thread resumptions.
suspensions	Pointer to destination for the total number of thread suspensions.
solicited_preemptions	Pointer to destination for the total number of thread preemptions as a result of a thread calling a ThreadX API service.
interrupt_preemptions	Pointer to destination for the total number of thread preemptions as a result of interrupt processing.
priority_inversions	Pointer to destination for the total number of thread priority inversions.
time_slices	Pointer to destination for the total number of thread time-slices.
relinquishes	Pointer to destination for the total number of thread relinquishes.
timeouts	Pointer to destination for the total number of thread suspension timeouts.
wait_aborts	Pointer to destination for the total number of thread wait aborts.
non_idle_returns	Pointer to destination for the number of times a thread returns to the system when another thread is ready to execute.
idle_returns	Pointer to destination for the number of times a thread returns to the system when no other thread is ready to execute (idle system).

> **NOTE:**
>
> Supplying a TX_NULL for any parameter indicates that the parameter is not required.

Return Values

TX_SUCCESS	(0x00)	Successful thread system performance get.
TX_FEATURE_NOT_ENABLED	(0xFF)	The system was not compiled with performance information enabled.

Allowed From

Initialization, threads, timers, and ISRs

Example

```
ULONG resumptions;
ULONG suspensions;
ULONG solicited_preemptions;
ULONG interrupt_preemptions;
ULONG priority_inversions;
ULONG time_slices;
ULONG relinquishes;
ULONG timeouts;
ULONG wait_aborts;
ULONG non_idle_returns;
ULONG idle_returns;
...
/* Retrieve performance information on all previously created
thread. */
status = tx_thread_performance_system_info_get(&resumptions,
                                &suspensions,
                                &solicited_preemptions,
                                &interrupt_preemptions,
                                &priority_inversions,
                                &time_slices, &relinquishes,
                                &timeouts,
                                &wait_aborts,
                                &non_idle_returns,
                                &idle_returns);
```

```
/* If status is TX_SUCCESS the performance information was
successfully retrieved. */
```

See Also

tx_thread_create, tx_thread_delete, tx_thread_entry_exit_notify, tx_thread_identify, tx_thread_info_get, tx_thread_performance_info_get, tx_thread_preemption_change, tx_thread_priority_change, tx_thread_relinquish, tx_thread_reset, tx_thread_resume, tx_thread_sleep, tx_thread_stack_error_notify, tx_thread_suspend, tx_thread_terminate, tx_thread_time_slice_change, tx_thread_wait_abort

tx_thread_preemption_change

Change preemption-threshold of application thread

Prototype

```
UINT tx_thread_preemption_change(TX_THREAD *thread_ptr,
  UINT new_threshold, UINT *old_threshold)
```

Description

This service changes the preemption-threshold of the specified thread. The preemption-threshold prevents preemption of the specified thread by threads whose priority is equal to or less than the preemption-threshold value. This service modifies the Thread Control Block through the parameter thread_ptr.

NOTE:

Using preemption-threshold disables time-slicing for the specified thread.

Input Parameters

thread_ptr	Pointer to a previously created application thread's Control Block.
new_threshold	New preemption-threshold priority level (0-31).

Output Parameter

old_threshold   Pointer to a location to return the previous preemption-threshold.

Return Values

TX_SUCCESS[6]   (0x00)   Successful priority change.
TX_THREAD_ERROR   (0x0E)   Invalid application thread pointer.
TX_PRIORITY_ERROR   (0x0F)   Specified new priority is not valid (a value other
                                                than 0-31).
TX_PTR_ERROR   (0x03)   Invalid pointer to previous priority storage
                                                location.
TX_CALLER_ERROR   (0x13)   Invalid caller of this service.

Allowed From

Threads and timers

Preemption Possible

Yes

Example

```
TX_THREAD my_thread;
UINT my_old_threshold;
UINT status;
...
/* Disable all preemption of the specified thread. The current
preemption-threshold is returned in "my_old_threshold". Assume
that "my_thread" has already been created. */
status = tx_thread_preemption_change(&my_thread, 0,
                                     &my_old_threshold);
/* If status equals TX_SUCCESS, the application thread is
non-preemptable by another thread. Note that ISRs are not
prevented by preemption disabling. */
```

---

[6]This value is not affected by the TX_DISABLE_ERROR_CHECKING define that is used to disable API error checking.

tx_thread_priority_change

Change priority of an application thread

Prototype

```
UINT tx_thread_priority_change(TX_THREAD *thread_ptr,
                                UINT new_priority,
                                UINT *old_priority)
```

Description

This service changes the priority of the specified thread. Valid priorities range from 0 to 31 (inclusive), where 0 represents the highest-priority level. This service modifies the Thread Control Block through the parameter thread_ptr.

NOTE:

The preemption-threshold of the specified thread is set automatically to the new priority. If you want to give the thread a different preemption-threshold, you must call the tx_thread_preemption_change service after this call.

Input Parameters

thread_ptr   Pointer to a previously created application thread's Control Block.
new_priority   New thread priority level (0-31).

Output Parameter

old_priority   Pointer to a location to return the thread's previous priority.

Return Values

TX_SUCCESS[7]   (0x00)   Successful priority change.
TX_THREAD_ERROR   (0x0E)   Invalid application thread pointer.

---

[7]This value is not affected by the TX_DISABLE_ERROR_CHECKING define that is used to disable API error checking.

TX_PRIORITY_ERROR	(0x0F)	Specified new priority is not valid (a value other than 0-31).
TX_PTR_ERROR	(0x03)	Invalid pointer to previous priority storage location.
TX_CALLER_ERROR	(0x13)	Invalid caller of this service.

Allowed From

Threads and timers

Preemption Possible

Yes

Example

```
TX_THREAD my_thread;
UINT my_old_priority;
UINT status;
...
/* Change the thread represented by "my_thread" to priority 0. */
status = tx_thread_priority_change(&my_thread, 0,
                                   &my_old_priority);
/* If status equals TX_SUCCESS, the application thread is now at
the highest priority level in the system. */
```

tx_thread_relinquish

Relinquish control to other application threads

Prototype

```
VOID tx_thread_relinquish(VOID)
```

Description

This service relinquishes processor control to other ready-to-run threads at the same or higher priority.

Parameters

None

Return Values

None

Allowed From

Threads

Preemption Possible

Yes

Example

```
ULONG run_counter_1 = 0;
ULONG run_counter_2 = 0;
...
/* Example of two threads relinquishing control to each other in
an infinite loop. Assume that both of these threads are ready and
have the same priority. The run counters will always stay within
one count of each other. */
VOID my_first_thread(ULONG thread_input)
{

  /* Endless loop of relinquish. */
  while(1)
  {
  /* Increment the run counter. */
  run_counter_1++;
  /* Relinquish control to other thread. */
  tx_thread_relinquish();
  }
}
VOID my_second_thread(ULONG thread_input)
{

  /* Endless loop of relinquish. */
```

```
while(1)
{
/* Increment the run counter. */
run_counter_2++;
/* Relinquish control to other thread. */
tx_thread_relinquish();
}
}
```

tx_thread_reset

Reset thread

Prototype

```
UINT tx_thread_reset(TX_THREAD *thread_ptr);
```

Description

This service resets the specified thread to execute at the entry point defined at thread creation. The thread must be in either a TX_COMPLETED or TX_TERMINATED state for it to be reset

> **NOTE:**
>
> The thread must be resumed for it to execute again.

Input Parameters

thread_ptr   Pointer to a previously created thread.

Return Values

TX_SUCCESS	(0x00)	Successful thread reset.
TX_NOT_DONE	(0x20)	Specified thread is not in a TX_COMPLETED or TX_TERMINATED state.
TX_THREAD_ERROR	(0x0E)	Invalid thread pointer.
TX_CALLER_ERROR	(0x13)	Invalid caller of this service.

Allowed From

Threads

Example

```
TX_THREAD my_thread;
...
/* Reset the previously created thread "my_thread." */
status = tx_thread_reset(&my_thread);
/* If status is TX_SUCCESS the thread is reset. */
```

See Also

tx_thread_create, tx_thread_delete, tx_thread_entry_exit_notify, tx_thread_identify, tx_thread_info_get, tx_thread_performance_info_get, tx_thread_preformance_system_info_get, tx_thread_preemption_change, tx_thread_priority_change, tx_thread_relinquish, tx_thread_resume, tx_thread_sleep, tx_thread_stack_error_notify, tx_thread_suspend, tx_thread_terminate, tx_thread_time_slice_change, tx_thread_wait_abort

tx_thread_resume

Resume suspended application thread

Prototype

```
UINT tx_thread_resume(TX_THREAD *thread_ptr)
```

Description

This service resumes or prepares for execution a thread that was previously suspended by a *tx_thread_suspend* call. In addition, this service resumes threads that were created without an automatic start.

Input Parameter

thread_ptr   Pointer to a suspended application thread's Control Block.

Return Values

TX_SUCCESS[8]	(0x00)	Successful resumption of thread.
TX_SUSPEND_LIFTED	(0x19)	Previously set delayed suspension was lifted.
TX_THREAD_ERROR	(0x0E)	Invalid application thread pointer.
TX_RESUME_ERROR	(0x12)	Specified thread is not suspended or was previously suspended by a service other than *tx_thread_suspend.*

Allowed From

Initialization, threads, timers, and ISRs

Preemption Possible

Yes

Example

```
TX_THREAD my_thread;
UINT status;
...
/* Resume the thread represented by "my_thread". */
status = tx_thread_resume(&my_thread);
/* If status equals TX_SUCCESS, the application thread is now
ready to execute. */
```

tx_thread_sleep

Suspend current thread for specified time.

Prototype

```
UINT tx_thread_sleep(ULONG timer_ticks)
```

---

[8]This value is not affected by the TX_DISABLE_ERROR_CHECKING define that is used to disable API error checking.

Description

This service causes the calling thread to suspend for the specified number of timer-ticks. The amount of physical time associated with a timer-tick is application-specific. This service can be called only from an application thread.

Input Parameter

timer_ticks   The number of timer-ticks to suspend the calling application thread, ranging from 0 to 0xFFFFFFFF (inclusive). If 0 is specified, the service returns immediately.

Return Values

TX_SUCCESS[9]	(0x00)	Successful thread sleep.
TX_WAIT_ABORTED	(0x1A)	Suspension was aborted by another thread, timer, or ISR.
TX_CALLER_ERROR	(0x13)	Service called from a non-thread.

Allowed From

Threads

Preemption Possible

Yes

Example

```
UINT status;
...
/* Make the calling thread sleep for 100 timer-ticks. */
status = tx_thread_sleep(100);
/* If status equals TX_SUCCESS, the currently running
application thread slept for the specified number of
timer-ticks. */
```

---

[9]This value is not affected by the TX_DISABLE_ERROR_CHECKING define that is used to disable API error checking.

tx_thread_stack_error_notify

Register thread stack error notification callback.

Prototype

```
UINT tx_thread_stack_error_notify(
                        VOID (*error_handler) (TX_THREAD *));
```

Description

This service registers a notification callback function for handling thread stack errors. When ThreadX detects a thread stack error during execution, it will call this notification function to process the error. Processing of the error is completely defined by the application. Anything from suspending the violating thread to resetting the entire system may be done.

NOTE:

The ThreadX library must be built with TX_ENABLE_STACK_CHECKING defined in order for this service to return performance information.

Input Parameters

error_handler   Pointer to application's stack error handling function. If this value is TX_NULL, the notification is disabled.

Return Values

TX_SUCCESS   (0x00)   Successful thread reset.
TX_FEATURE_NOT_ENABLED (0xFF)   The system was not compiled with performance information enabled.

Allowed From

Initialization, threads, timers, and ISRs

Example

```
void my_stack_error_handler(TX_THREAD *thread_ptr);
```

```
/* Register the "my_stack_error_handler" function with ThreadX so
that thread stack errors can be handled by the application. */
status = tx_thread_stack_error_notify(my_stack_error_handler);
/* If status is TX_SUCCESS the stack error handler is
registered.*/
```

See Also

tx_thread_create, tx_thread_delete, tx_thread_entry_exit_notify, tx_thread_identify, tx_thread_info_get, tx_thread_performance_info_get, tx_thread_preformance_system_info_get, tx_thread_preemption_change, tx_thread_priority_change, tx_thread_relinquish, tx_thread_reset, tx_thread_resume, tx_thread_sleep, tx_thread_suspend, tx_thread_terminate, tx_thread_time_slice_change, tx_thread_wait_abort

tx_thread_suspend

Suspend an application thread

Prototype

```
UINT tx_thread_suspend(TX_THREAD *thread_ptr)
```

Description

This service suspends the specified application thread. A thread may call this service to suspend itself.

NOTE:

If the specified thread is already suspended for another reason, this new suspension is held internally until the prior suspension is lifted. When the prior suspension is lifted, this new unconditional suspension of the specified thread is performed. Further unconditional suspension requests have no effect.

Once suspended, the thread must be resumed by *tx_thread_resume* in order to execute again.

Input Parameter

thread_ptr   Pointer to an application Thread Control Block.

Return Values

TX_SUCCESS[10]   (0x00)   Successful suspension of thread.
TX_THREAD_ERROR   (0x0E)   Invalid application thread pointer.
TX_SUSPEND_ERROR9 (0x14)   Specified thread is in a terminated or completed state.
TX_CALLER_ERROR   (0x13)   Invalid caller of this service.

Allowed From

Threads and timers

Preemption Possible

Yes

Example

```
TX_THREAD my_thread;
UINT status;
...
/* Suspend the thread represented by "my_thread". */
status = tx_thread_suspend(&my_thread);
/* If status equals TX_SUCCESS, the application thread is
unconditionally suspended. */
```

tx_thread_terminate

Terminates an application thread

Prototype

```
UINT tx_thread_terminate(TX_THREAD *thread_ptr)
```

---

[10]This value is not affected by the TX_DISABLE_ERROR_CHECKING define that is used to disable API error checking.

Description

This service terminates the specified application thread regardless of whether or not the thread is suspended. A thread may call this service to terminate itself. This service modifies the Thread Control Block through the parameter thread_ptr.

> **NOTE:**
> Once terminated, the thread must be deleted and re-created in order for it to execute again.

Input Parameter

thread_ptr   Pointer to a previously created application thread's Control Block.

Return Values

TX_SUCCESS[11]	(0x00)	Successful thread termination.
TX_THREAD_ERROR	(0x0E)	Invalid application thread pointer.
TX_CALLER_ERROR	(0x13)	Invalid caller of this service.

Allowed From

Threads and timers

Preemption Possible

Yes

Example

```
TX_THREAD my_thread;
UINT status;
…
/* Terminate the thread represented by "my_thread". */
status = tx_thread_terminate(&my_thread);
```

---

[11]This value is not affected by the TX_DISABLE_ERROR_CHECKING define that is used to disable API error checking.

```
/* If status equals TX_SUCCESS, the thread is terminated and
cannot execute again until it is deleted and re-created. */
```

tx_thread_time_slice_change

Changes time-slice of application thread

Prototype

```
UINT tx_thread_time_slice_change(TX_THREAD *thread_ptr,
                                 ULONG new_time_slice,
                                 ULONG *old_time_slice)
```

Description

This service changes the time-slice of the specified application thread. Selecting a time-slice for a thread ensures that it won't execute more than the specified number of timer-ticks before other threads of the same or higher priorities have a chance to execute. This service modifies the Thread Control Block through the parameter thread_ptr.

NOTE:

Using preemption-threshold disables time-slicing for the specified thread.

Input Parameters

thread_ptr	Pointer to a previously created application thread's Control Block.
new_time_slice	New time-slice value. Legal values include TX_NO_TIME_SLICE and numeric values from 1 to 0xFFFFFFFF (inclusive).

Output Parameter

old_time_slice	Pointer to location for storing the previous time-slice value of the specified thread.

Return Values

TX_SUCCESS[12]   (0x00)   Successful time-slice chance.
TX_THREAD_ERROR (0x0E)   Invalid application thread pointer.
TX_PTR_ERROR   (0x03)   Invalid pointer to previous time-slice storage location.
TX_CALLER_ERROR (0x13)   Invalid caller of this service.

Allowed From

Threads and timers

Preemption Possible

No

Example

```
TX_THREAD my_thread;
ULONG my_old_time_slice;
UINT status;
...
/* Change the time-slice of the thread associated with "my_thread"
to 20. This will mean that "my_thread" can only run for 20 timer-
ticks consecutively before other threads of equal or higher
priority get a chance to run. */
status = tx_thread_time_slice_change(&my_thread, 20,
                                     &my_old_time_slice);
/* If status equals TX_SUCCESS, the time-slice has been changed to
20 and the previous time-slice is in "my_old_time_slice." */
```

tx_thread_wait_abort

Abort suspension of specified thread

Prototype

```
UINT tx_thread_wait_abort(TX_THREAD *thread_ptr)
```

---

[12]This value is not affected by the TX_DISABLE_ERROR_CHECKING define that is used to disable API error checking.

Description

This service aborts a sleep or any other object's suspension of the specified thread. If the wait is aborted, a TX_WAIT_ABORTED value is returned from the service that the thread was waiting on.

> **NOTE:**
>
> This service does not release an unconditional suspension that was caused by the tx_thread_suspend service

Input Parameter

thread_ptr   Pointer to a previously created application thread's Control Block.

Return Values

TX_SUCCESS	(0x00)	Successful thread wait abort.
TX_THREAD_ERROR	(0x0E)	Invalid application thread pointer.
TX_WAIT_ABORT_ERROR	(0x1B)	Specified thread is not in a waiting state.

Allowed From

Initialization, threads, timers, and ISRs

Preemption Possible

Yes

Example

```
TX_THREAD my_thread;
UINT status;
...
/* Abort the suspension condition of "my_thread." */
status = tx_thread_wait_abort(&my_thread);
/* If status equals TX_SUCCESS, the thread is now ready again,
with a return value showing its suspension was aborted
(TX_WAIT_ABORTED). */
```

Internal System Clock Services

The time services described in this appendix are:

tx_time_get   Retrieves the current time
tx_time_set   Sets the current time

tx_time_get

Retrieves the current time

Prototype

```
ULONG tx_time_get(VOID)
```

Description

This service returns the contents of the internal system clock. Each timer-tick increases the internal system clock count by one. The system clock is set to zero during initialization and can be changed to a specific value by the service *tx_time_set*.

WARNING:

The physical time interval each timer-tick represents is application-specific.

Parameters

None

Return Values

system clock ticks   Value of the internal, free running, system clock.

Allowed From

Initialization, threads, timers, and ISRs

Preemption Possible

No

Example

```
ULONG current_time;
...
/* Pickup the current system time, in timer-ticks. */
current_time = tx_time_get();
/* Current time now contains a copy of the internal system clock. */
```

tx_time_set

Sets the current time

Prototype

```
VOID tx_time_set(ULONG new_time)
```

Description

This service sets the internal system clock to the specified value. Each subsequent timer-tick increases the internal system clock count by one.

WARNING:

The physical time interval each timer-tick represents is application-specific.

Return Values

None

Allowed From

Threads, timers, and ISRs

Preemption Possible

No

Example

```
/* Set the internal system time to 0x1234. */
tx_time_set(0x1234);
/* Current time now contains 0x1234 until the next timer interrupt. */
```

Application Timer Services

The application timer services described in this appendix include:

tx_timer_activate	Activate an application timer
tx_timer_change	Change an application timer
tx_timer_create	Create an application timer
tx_timer_deactivate	Deactivate an application timer
tx_timer_delete	Delete an application timer
tx_timer_info_get	Retrieve information about an application timer
tx_timer_performance_info_get	Get timer performance information
tx_timer_performance_system_info_get	Get timer system performance information

tx_timer_activate

Activate an application timer

Prototype

```
UINT tx_timer_activate (TX_TIMER*timer_ptr)
```

Description

This service activates the specified application timer. The expiration routines of timers that expire at the same time are executed in the order they were activated. This service modifies the Application Timer Control Block through the parameter timer_ptr.

Input Parameter

timer_ptr   Pointer to a previously created application timer's Control Block.

Return Values

TX_SUCCESS[1]	(0×00)	Successful application timer activation.
TX_TIMER_ERROR	(0×15)	Invalid application timer pointer.
TX_ACTIVATE_ERROR1	(0×17)	Timer was already active.

Allowed From

Initialization, threads, timers, and ISRs

Preemption Possible

No

Example

```
TX_TIMER my_timer;
UINT status;
…
/* Activate an application timer. Assume that the application
timer has already been created. */
status = tx_timer_activate (&my_timer);
/* If status equals TX_SUCCESS, the application timer is now
active. */
```

tx_timer_change

Change an application timer

Prototype

```
UINT tx_timer_change (TX_TIMER *timer_ptr, ULONG initial_ticks,
                      ULONG reschedule_ticks)
```

---

[1]This value is not affected by the TX_DISABLE_ERROR_CHECKING define that is used to disable API error checking.

Description

This service changes the expiration characteristics of the specified application timer. The timer must be deactivated prior to calling this service. This service modifies the Application Timer Control Block through the parameter timer_ptr.

WARNING:

You must call the tx_timer_activate service after calling this service in order to restart the timer.

Input Parameters

timer_ptr	Pointer to a previously created timer's Control Block.
initial_ticks	Specifies the initial number of timer-ticks for timer expiration. Legal values range from 1 to 0×FFFFFFFF (inclusive).
reschedule_ticks	Specifies the number of timer-ticks for all timer expirations after the first. A zero for this parameter makes the timer a one-shot timer. Otherwise, for periodic timers, legal values range from 1 to 0×FFFFFFFF (inclusive).

Return Values

TX_SUCCESS[2]	(0×00)	Successful application timer change.
TX_TIMER_ERROR	(0×15)	Invalid application timer pointer.
TX_TICK_ERROR	(0×16)	Invalid value (a zero) supplied for initial timer-ticks.
TX_CALLER_ERROR	(0×13)	Invalid caller of this service.

Allowed From

Threads, timers, and ISRs

Preemption Possible

No

---

[2]This value is not affected by the TX_DISABLE_ERROR_CHECKING define that is used to disable API error checking.

Example

```
TX_TIMER my_timer;
UINT status;
...
/* Change a previously created and now deactivated timer to expire
every 50 timer-ticks for initial and subsequent expirations. */
status = tx_timer_change (&my_timer,50, 50);
/* If status equals TX_SUCCESS, the specified timer is changed to
expire every 50 timer-ticks. */
/* Activate the specified timer to get it started again. */
status = tx_timer_activate(&my_timer);
```

tx_timer_create

Create an application timer

Prototype

```
UINT tx_timer_create (TX_TIMER *timer_ptr, CHAR *name_ptr,

        VOID (*expiration_function)(ULONG),
        ULONG expiration_input, ULONG initial_ticks,
        ULONG reschedule_ticks, UINT auto_activate)
```

Description

This service creates an application timer with the specified expiration function and periodic expiration interval. This service initializes the timer Control Block through the parameter timer_ptr.

Input Parameters

timer_ptr	Pointer to a timer Control Block
name_ptr	Pointer to the name of the timer.
expiration_function	Application function to call when the timer expires.
expiration_input	Input to pass to expiration function when the timer expires.
initial_ticks	Specifies the initial number of timer-ticks for timer expiration. Legal values range from 1 to 0×FFFFFFFF (inclusive).

reschedule_ticks   Specifies the number of timer-ticks for all timer expirations after the first. A zero for this parameter makes the timer a one-shot timer. Otherwise, for periodic timers, legal values range from 1 to 0×FFFFFFFF (inclusive).

Auto_activate   Determines whether the timer is automatically activated during creation. If this value is TX_AUTO_ACTIVATE (0 × 01) the timer is made active. Otherwise, if the value TX_NO_ACTIVATE (0×00) is selected, the timer is created in a non-active state. In this case, a subsequent tx_timer_activate service call is necessary to get the timer actually started.

Return Values

TX_SUCCESS[3]	(0×00)	Successful application timer creation.
TX_TIMER_ERROR	(0×15)	Invalid application timer pointer. Either the pointer is NULL or the timer has already been created.
TX_TICK_ERROR	(0×16)	Invalid value (a zero) supplied for initial timer-ticks.
TX_ACTIVATE_ERROR	(0×17)	Invalid activation selected.
TX_CALLER_ERROR	(0×13)	Invalid caller of this service.

Allowed From

Initialization and threads

Preemption Possible

No

Example

```
TX_TIMER my_timer;
UINT status;
```

---

[3]This value is not affected by the TX_DISABLE_ERROR_CHECKING define that is used to disable API error checking.

```
...
/* Create an application timer that executes "my_timer_function"
after 100 timer-ticks initially and then after every 25 timer-
ticks. This timer is specified to start immediately! */
status = tx_timer_create (&my_timer,"my_timer_name",
                          my_timer_function,
                          0x1234, 100, 25, TX_AUTO_ACTIVATE);
/* If status equals TX_SUCCESS, my_timer_function will be called
100 timer-ticks later and then called every 25 timer-ticks. Note
that the value 0x1234 is passed to my_timer_function every time it
is called. */
```

tx_timer_deactivate

Deactivate an application timer

Prototype

```
UINT tx_timer_deactivate (TX_TIMER *timer_ptr)
```

Description

This service deactivates the specified application timer. If the timer is already deactivated, this service has no effect. This service may modify the timer Control Block through the parameter timer_ptr.

Input Parameter

timer_ptr   Pointer to a previously created application timer's Control Block.

Return Values

TX_SUCCESS[4]	(0×00)	Successful application timer deactivation.
TX_TIMER_ERROR	(0×15)	Invalid application timer pointer.

---

[4]This value is not affected by the TX_DISABLE_ERROR_CHECKING define that is used to disable API error checking.

Allowed From

Initialization, threads, timers, and ISRs

Preemption Possible

No

Example

```
TX_TIMER my_timer;
UINT status;
...
/* Deactivate an application timer. Assume that the application
timer has already been created. */
status = tx_timer_deactivate(&my_timer);
/* If status equals TX_SUCCESS, the application timer is now
deactivated. */
```

tx_timer_delete

Delete an application timer

Prototype

```
UINT tx_timer_delete(TX_TIMER *timer_ptr)
```

Description

This service deletes the specified application timer.

WARNING:

It is the application's responsibility to prevent use of a deleted timer.

Input Parameter

timer_ptr   Pointer to a previously created application timer's Control Block.

Return Values

TX_SUCCESS[5]	(0×00)	Successful application timer deletion.
TX_TIMER_ERROR	(0×15)	Invalid application timer pointer.
TX_CALLER_ERROR	(0×13)	Invalid caller of this service.

Allowed From

Threads

Preemption Possible

No

Example

```
TX_TIMER my_timer;
UINT status;
…
/* Delete application timer. Assume that the application timer has
already been created. */
status = tx_timer_delete(&my_timer);
/* If status equals TX_SUCCESS, the application timer is deleted.
*/
```

tx_timer_info_get

Retrieve information about an application timer

Prototype

```
UINT tx_timer_info_get(TX_TIMER *timer_ptr, CHAR **name,
                       UINT *active, ULONG
                       *remaining_ticks,
                       ULONG *reschedule_ticks,
                       TX_TIMER **next_timer)
```

---

[5]This value is not affected by the TX_DISABLE_ERROR_CHECKING define that is used to disable API error checking.

Description

This service retrieves information about the specified application timer.

Input Parameter

timer_ptr   Pointer to a previously created application timer's Control Block.

Output Parameters

name	Pointer to destination for the pointer to the timer's name.
active	Pointer to destination for the timer active indication. If the timer is inactive or this service is called from the timer itself, a TX_FALSE value is returned. Otherwise, if the timer is active, a TX_TRUE value is returned.
remaining_ticks	Pointer to destination for the number of timer-ticks left before the timer expires.
reschedule_ticks	Pointer to destination for the number of timer-ticks that will be used to automatically reschedule this timer. If the value is zero, then the timer is a one-shot and won't be rescheduled.
next_timer	Pointer to destination for the pointer of the next created application timer.

Return Values

TX_SUCCESS[6]	(0×00)	Successful timer information retrieval.
TX_TIMER_ERROR	(0×15)	Invalid application timer pointer.
TX_PTR_ERROR	(0×03)	Invalid pointer (NULL) for any destination pointer.

Allowed From

Initialization, threads, timers, and ISRs

---

[6]This value is not affected by the TX_DISABLE_ERROR_CHECKING define that is used to disable API error checking.

Preemption Possible

No

Example

```
TX_TIMER my_timer;
CHAR *name;
UINT active;
ULONG remaining_ticks;
ULONG reschedule_ticks;
TX_TIMER *next_timer;
UINT status;
...
/* Retrieve information about the previously created application
timer "my_timer." */
status = tx_timer_info_get(&my_timer, &name, &active,
        &remaining_ticks, &reschedule_ticks, &next_timer);
/* If status equals TX_SUCCESS, the information requested is
valid. */
```

tx_timer_performance_info_get

Get timer performance information

Prototype

```
UINT tx_timer_performance_info_get(TX_TIMER *timer_ptr,
                            ULONG *activates,
                            ULONG *reactivates,
                            ULONG *deactivates,
                            ULONG *expirations,
                            ULONG *expiration_adjusts);
```

Description

This service retrieves performance information about the specified application timer.

> **NOTE:**
>
> The ThreadX library and application must be built with TX_TIMER_ENABLE_
> PERFORMANCE_INFO defined for this service to return performance information.

Input Parameters

timer_ptr	Pointer to previously created timer.
activates	Pointer to destination for the number of activation requests performed on this timer.
reactivates	Pointer to destination for the number of automatic reactivations performed on this periodic timer.
deactivates	Pointer to destination for the number of deactivation requests performed on this timer.
expirations	Pointer to destination for the number of expirations of this timer.
expiration_adjusts	Pointer to destination for the number of internal expiration adjustments performed on this timer. These adjustments are done in the timer interrupt processing for timers that are larger than the default timer list size (by default timers with expirations greater than 32 ticks).

> **NOTE:**
>
> Supplying a TX_NULL for any parameter indicates the parameter is not required.

Return Values

TX_SUCCESS	(0x00)	Successful timer performance get.
TX_PTR_ERROR	(0x03)	Invalid timer pointer.
TX_FEATURE_NOT_ENABLED	(0xFF)	The system was not compiled with performance information enabled.

Allowed From

Initialization, threads, timers, and ISRs

Example

```
TX_TIMER my_timer;
ULONG activates;
ULONG reactivates;
ULONG deactivates;
ULONG expirations;
ULONG expiration_adjusts;
...
/* Retrieve performance information on the previously created
timer. */
status = tx_timer_performance_info_get(&my_timer, &activates,
                                       &reactivates,&deactivates,
                                       &expirations,
                                       &expiration_adjusts);
/* If status is TX_SUCCESS the performance information was
successfully retrieved. */
```

See Also

tx_timer_activate, tx_timer_change, tx_timer_create, tx_timer_deactivate, tx_timer_
delete, tx_timer_info_get, tx_timer_performance_system_info_get

tx_timer_performance_system_info_get

Get timer system performance information

Prototype

```
UINT tx_timer_performance_system_info_get(ULONG *activates,
                                      ULONG *reactivates,
                                      ULONG *deactivates,
                                      ULONG *expirations,
                                      ULONG *expiration_adjusts);
```

Description

This service retrieves performance information about all the application timers in the
system.

> **NOTE:**
>
> The ThreadX library and application must be built with TX_TIMER_ENABLE_
> PERFORMANCE_INFO defined for this service to return performance information.

Input Parameters

activates	Pointer to destination for the total number of activation requests performed on all timers.
reactivates	Pointer to destination for the total number of automatic reactivation performed on all periodic timers.
deactivates	Pointer to destination for the total number of deactivation requests performed on all timers.
expirations	Pointer to destination for the total number of expirations on all timers.
expiration_adjusts	Pointer to destination for the total number of internal expiration adjustments performed on all timers. These adjustments are done in the timer interrupt processing for timers that are larger than the default timer list size (by default timers with expirations greater than 32 ticks).

> **NOTE:**
>
> Supplying a TX_NULL for any parameter indicates that the parameter is not
> required.

Return Values

TX_SUCCESS	(0x00)	Successful timer system performance get.
TX_FEATURE_NOT_ENABLED	(0xFF)	The system was not compiled with performance information enabled.

Allowed From

Initialization, threads, timers, and ISRs

Example

```
ULONG activates;
ULONG reactivates;
ULONG deactivates;
ULONG expirations;
ULONG expiration_adjusts;
...
/* Retrieve performance information on all previously created
timers. */
status = tx_timer_performance_system_info_get(&activates,
                                    &reactivates,
                                    &deactivates, &expirations,
                                    &expiration_adjusts);
/* If status is TX_SUCCESS the performance information was
successfully retrieved. */
```

See Also

tx_timer_activate, tx_timer_change, tx_timer_create, tx_timer_deactivate, tx_timer_delete, tx_
timer_info_get, tx_timer_performance_info_get

ThreadX API[1]

ThreadX Entry

```
VOID tx_kernel_enter(VOID);
```

Memory Byte Pool Services

```
UINT tx_byte_allocate(TX_BYTE_POOL *pool_ptr,
  VOID **memory_ptr,
  ULONG memory_size,
  ULONG wait_option);

UINT tx_byte_pool_create(TX_BYTE_POOL *pool_ptr,
  CHAR *name_ptr,
  VOID *pool_start,
  ULONG pool_size);

UINT tx_byte_pool_delete(TX_BYTE_POOL *pool_ptr);

UINT tx_byte_pool_info_get(TX_BYTE_POOL *pool_ptr,
  CHAR **name,
  ULONG *available_bytes,
  ULONG *fragments,
  TX_THREAD **first_suspended,
  ULONG *suspended_count,
  TX_BYTE_POOL **next_pool);
```

---

[1] ThreadX is a registered trademark of Express Logic, Inc. The ThreadX API, associated data structures, and data types are copyrights of Express Logic, Inc.

```
UINT tx_byte_pool_performance_info_get(TX_BYTE_POOL *pool_ptr,
  ULONG *allocates,
  ULONG *releases,
  ULONG *fragments_searched,
  ULONG *merges,
  ULONG *splits,
  ULONG *suspensions,
  ULONG *timeouts);

UINT tx_byte_pool_performance_system_info_get(ULONG *allocates,
  ULONG *releases,
  ULONG *fragments_searched,
  ULONG *merges,
  ULONG *splits,
  ULONG *suspensions,
  ULONG *timeouts);

UINT tx_byte_pool_prioritize(TX_BYTE_POOL *pool_ptr);

UINT tx_byte_release(VOID *memory_ptr);
```

Memory Block Pool Services

```
UINT tx_block_allocate(TX_BLOCK_POOL *pool_ptr,
  VOID **block_ptr,
  ULONG wait_option);

UINT tx_block_pool_create(TX_BLOCK_POOL *pool_ptr,
  CHAR *name_ptr,
  ULONG block_size,
  VOID *pool_start,
  ULONG pool_size);

UINT tx_block_pool_delete(TX_BLOCK_POOL *pool_ptr);

UINT tx_block_pool_info_get(TX_BLOCK_POOL *pool_ptr,
  CHAR **name,
  ULONG *available_blocks,
  ULONG *total_blocks,
  TX_THREAD **first_suspended,
  ULONG *suspended_count,
  TX_BLOCK_POOL **next_pool);
```

```
UINT tx_block_pool_performance_info_get(TX_BLOCK_POOL *pool_ptr,
  ULONG *allocates,
  ULONG *releases,
  ULONG *suspensions,
  ULONG *timeouts);

UINT tx_block_pool_performance_system_info_get(ULONG *allocates,
  ULONG *releases,
  ULONG *suspensions,
  ULONG *timeouts);

UINT tx_block_pool_prioritize(TX_BLOCK_POOL *pool_ptr);

UINT tx_block_release(VOID *block_ptr);
```

Event Flag Services

```
UINT tx_event_flags_create(TX_EVENT_FLAGS_GROUP *group_ptr,
  CHAR *name_ptr);

UINT tx_event_flags_delete(TX_EVENT_FLAGS_GROUP *group_ptr);

UINT tx_event_flags_get(TX_EVENT_FLAGS_GROUP *group_ptr,
  ULONG requested_flags,
  UINT get_option,
  ULONG *actual_flags_ptr,
  ULONG wait_option);

UINT tx_event_flags_info_get(TX_EVENT_FLAGS_GROUP *group_ptr,
  CHAR **name,
  ULONG *current_flags,
  TX_THREAD **first_suspended,
  ULONG *suspended_count,
  TX_EVENT_FLAGS_GROUP **next_group);

UINT tx_event_flags_performance_info_get(TX_EVENT_FLAGS_GROUP
  *group_ptr,
  ULONG *sets,
  ULONG *gets,
  ULONG *suspensions,
  ULONG *timeouts);
```

```
UINT tx_event_flags_performance_system_info_get(ULONG *sets,
  ULONG *gets,ULONG *suspensions,
  ULONG *timeouts);

UINT tx_event_flags_set(TX_EVENT_FLAGS_GROUP *group_ptr,
  ULONG flags_to_set,
  UINT set_option);

UINT tx_event_flags_set_notify(TX_EVENT_FLAGS_GROUP *group_ptr,
  VOID(*events_set_notify)(TX_EVENT_FLAGS_GROUP *));
```

Interrupt Control Service

```
UINT tx_interrupt_control(UINT new_posture);
```

Message Queue Services

```
UINT tx_queue_create(TX_QUEUE *queue_ptr,
  CHAR *name_ptr,
  UINT message_size,
  VOID *queue_start,
  ULONG queue_size);

UINT tx_queue_delete(TX_QUEUE *queue_ptr);

UINT tx_queue_flush(TX_QUEUE *queue_ptr);

UINT tx_queue_front_send(TX_QUEUE *queue_ptr,
  VOID *source_ptr,
  ULONG wait_option);

UINT tx_queue_info_get(TX_QUEUE *queue_ptr,
  CHAR **name,
  ULONG *enqueued,
  ULONG *available_storage,
  TX_THREAD **first_suspended,
  ULONG *suspended_count,
  TX_QUEUE **next_queue);

UINT tx_queue_performance_info_get(TX_QUEUE *queue_ptr,
  ULONG *messages_sent,
  ULONG *messages_received,
```

```
    ULONG *empty_suspensions,
    ULONG *full_suspensions,
    ULONG *full_errors,
    ULONG *timeouts);

UINT tx_queue_performance_system_info_get(ULONG *messages_sent,
    ULONG *messages_received,
    ULONG *empty_suspensions,
    ULONG *full_suspensions,
    ULONG *full_errors,
    ULONG *timeouts);

UINT tx_queue_prioritize(TX_QUEUE *queue_ptr);

UINT tx_queue_receive(TX_QUEUE *queue_ptr,
    VOID *destination_ptr,
    ULONG wait_option);

UINT tx_queue_send(TX_QUEUE *queue_ptr,
    VOID *source_ptr,
    ULONG wait_option);

UINT tx_queue_send_notify(TX_QUEUE *queue_ptr,
    VOID (*queue_send_notify)(TX_QUEUE *));
```

Semaphore Services

```
UINT tx_semaphore_ceiling_put(TX_SEMAPHORE *semaphore_ptr,
    ULONG ceiling);

UINT tx_semaphore_create(TX_SEMAPHORE *semaphore_ptr,
    CHAR *name_ptr,
    ULONG initial_count);

UINT tx_semaphore_delete(TX_SEMAPHORE *semaphore_ptr);

UINT tx_semaphore_get(TX_SEMAPHORE *semaphore_ptr,
    ULONG wait_option);

UINT tx_semaphore_info_get(TX_SEMAPHORE *semaphore_ptr,
    CHAR **name,
    ULONG *current_value,
    TX_THREAD **first_suspended,
```

```
    ULONG *suspended_count,
    TX_SEMAPHORE **next_semaphore);

UINT tx_semaphore_performance_info_get(TX_SEMAPHORE *semaphore_ptr,
    ULONG *puts,
    ULONG *gets,
    ULONG *suspensions,
    ULONG *timeouts);

UINT tx_semaphore_performance_system_info_get(ULONG *puts,
    ULONG *gets,
    ULONG *suspensions,
    ULONG *timeouts);

UINT tx_semaphore_prioritize(TX_SEMAPHORE *semaphore_ptr);

UINT tx_semaphore_put(TX_SEMAPHORE *semaphore_ptr);

UINT tx_semaphore_put_notify(TX_SEMAPHORE *semaphore_ptr,
    VOID (*semaphore_put_notify)(TX_SEMAPHORE *));
```

Mutex Services

```
UINT tx_mutex_create(TX_MUTEX *mutex_ptr,
    CHAR *name_ptr,
    UINT inherit);

UINT tx_mutex_delete(TX_MUTEX *mutex_ptr);

UINT tx_mutex_get(TX_MUTEX *mutex_ptr,
    ULONG wait_option);

UINT tx_mutex_info_get(TX_MUTEX *mutex_ptr,
    CHAR **name,
    ULONG *count,
    TX_THREAD **owner,
    TX_THREAD **first_suspended,
    ULONG *suspended_count,
    TX_MUTEX **next_mutex);

UINT tx_mutex_performance_info_get(TX_MUTEX *mutex_ptr,
    ULONG *puts,
    ULONG *gets,
    ULONG *suspensions,
```

```
  ULONG *timeouts,
  ULONG *inversions,
  ULONG *inheritances);

UINT tx_mutex_performance_system_info_get(ULONG *puts,
  ULONG *gets,
  ULONG *suspensions,
  ULONG *timeouts,
  ULONG *inversions,
  ULONG *inheritances);

UINT tx_mutex_prioritize(TX_MUTEX *mutex_ptr);

UINT tx_mutex_put(TX_MUTEX *mutex_ptr);
```

Thread Services

```
UINT tx_thread_create(TX_THREAD *thread_ptr,
  CHAR *name_ptr,
  VOID(*entry_function)(ULONG),
  ULONG entry_input,
  VOID *stack_start,
  ULONG stack_size,
  UINT priority,
  UINT preempt_threshold,
  ULONG time_slice,
  UINT auto_start);

UINT tx_thread_delete(TX_THREAD *thread_ptr);

UINT tx_thread_entry_exit_notify(TX_THREAD *thread_ptr,
  VOID(*entry_exit_notify)(TX_THREAD *, UINT));

TX_THREAD *tx_thread_identify(VOID);

UINT tx_thread_info_get(TX_THREAD *thread_ptr,
  CHAR **name,
  UINT *state,
  ULONG *run_count,
  UINT *priority,
  UINT *preemption_threshold,
  ULONG *time_slice,
```

```
  TX_THREAD **next_thread,
  TX_THREAD **next_suspended_thread);

UINT tx_thread_performance_info_get(TX_THREAD *thread_ptr,
  ULONG *resumptions,
  ULONG *suspensions,
  ULONG *solicited_preemptions,
  ULONG *interrupt_preemptions,
  ULONG *priority_inversions,
  ULONG *time_slices,
  ULONG *relinquishes,
  ULONG *timeouts,
  ULONG *wait_aborts,
  TX_THREAD **last_preempted_by);

UINT tx_thread_performance_system_info_get(ULONG *resumptions,
  ULONG *suspensions,
  ULONG *solicited_preemptions,
  ULONG *interrupt_preemptions,
  ULONG *priority_inversions,
  ULONG *time_slices,
  ULONG *relinquishes,
  ULONG *timeouts,
  ULONG *wait_aborts,
  ULONG *non_idle_returns,
  ULONG *idle_returns);

UINT tx_thread_preemption_change(TX_THREAD *thread_ptr,
  UINT new_threshold,
  UINT *old_threshold);

UINT tx_thread_priority_change(TX_THREAD *thread_ptr,
  UINT new_priority,
  UINT *old_priority);

VOID tx_thread_relinquish(VOID);

UINT tx_thread_reset(TX_THREAD *thread_ptr);

UINT tx_thread_resume(TX_THREAD *thread_ptr);

UINT tx_thread_sleep(ULONG timer_ticks);

UINT tx_thread_stack_error_notify(VOID(*error_handler)
  (TX_THREAD *));
```

UINT tx_thread_suspend(TX_THREAD *thread_ptr);

UINT tx_thread_terminate(TX_THREAD *thread_ptr);

UINT tx_thread_time_slice_change(TX_THREAD *thread_ptr,
  ULONG new_time_slice,
  ULONG *old_time_slice);

UINT tx_thread_wait_abort(TX_THREAD *thread_ptr);

Time Services

ULONG tx_time_get(VOID);

VOID tx_time_set(ULONG new_time);

Timer Services

UINT tx_timer_activate(TX_TIMER *timer_ptr);

UINT tx_timer_change(TX_TIMER *timer_ptr,
  ULONG initial_ticks,
  ULONG reschedule_ticks);

UINT tx_timer_create(TX_TIMER *timer_ptr,
  CHAR *name_ptr,
  VOID(*expiration_function)(ULONG),
  ULONG expiration_input,
  ULONG initial_ticks,
  ULONG reschedule_ticks,
  UINT auto_activate);

UINT tx_timer_deactivate(TX_TIMER *timer_ptr);

UINT tx_timer_delete(TX_TIMER *timer_ptr);

UINT tx_timer_info_get(TX_TIMER *timer_ptr,
  CHAR **name,
  UINT *active,
  ULONG *remaining_ticks,
  ULONG *reschedule_ticks,
  TX_TIMER **next_timer);

```
UINT tx_timer_performance_info_get(TX_TIMER *timer_ptr,
  ULONG *activates,
  ULONG *reactivates,
  ULONG *deactivates,
  ULONG *expirations,
  ULONG *expiration_adjusts);

UINT tx_timer_performance_system_info_get(ULONG *activates,
  ULONG *reactivates,
  ULONG *deactivates,
  ULONG *expirations,
  ULONG *expiration_adjusts);
```

Index